Introduction to the Fine Structure of Plant Cells

Myron C. Ledbetter · Keith R. Porter

Introduction to the Fine Structure of Plant Cells

With 51 Plates and 8 Text Figures

Springer-Verlag
Berlin · Heidelberg · New York 1970

Dr. Myron C. Ledbetter
Brookhaven National Laboratories
Upton, L.I., NY 11973, U.S.A.

Professor Keith R. Porter
Harvard University
Cambridge, MA 02138, U.S.A.
and
The University of Colorado
Boulder, CO 80302, U.S.A.

ISBN 978-3-642-87139-9 ISBN 978-3-642-87137-5 (eBook)
DOI 10.1007/978-3-642-87137-5

Softcover reprint of the hardcover 1st edition 1970

Typesetting, printing and binding: Universitätsdruckerei H. Stürtz AG, Würzburg. Blockmaker: Gustav Dreher, Württemb. Graphische Kunstanstalt GmbH, Stuttgart.

Preface

It is appropiate to the contents of this book to recall a few highlights in the history of plant cytology from its inception over three centuries ago. Robert Hooke in 1663 presented his observations of what he called "cells" in cork and other plant parts and beautifully illustrated and described these in his classic "Micrographia" published two years later. More detailed exploration of the cell and its contents awaited almost two centuries for Robert Brown's discovery of the nucleus in 1831. Discoveries of other cell organelles followed, particularly in the latter part of the 19th and early part of this century. As is frequently noted each of these achievements was preceeded by advances in the resolution of the microscope. Now history repeats and recent developments in electron microscopy have given the biologist the opportunity to study cell morphology in far greater detail than at any time previously. Indeed, the resolution of the electron microscope is several hundredfold better than that available in the finest light microscopes. These advances in instrumentation plus improvements in the techniques of specimen preparation have made possible the examination of plant cells of almost any type. It is the resulting wealth of new information now accessible to the botanical cytologist that has prompted this publication.

In this book we have brought together electron micrographs representing a number of cell types from higher plants. Each plate is accompanied by a description to introduce the reader to the salient features of cell morphology. The micrographs have been freely interpreted as to the relation of stucture and function, the better to provoke thought on the part of the student. Markers have been placed on the micrographs to lead the eye more easily to structures of interest. Finally, selected references are provided with each plate to guide the reader to examples of the available literature. It is hoped that this atlas will serve to introduce the professional botanist as well as the advanced student to this rapidly progressing area of plant research.

Most of the micrographs reproduced here originated in the Laboratory for Cell Biology at Harvard University and have not been published elsewhere. The balance were taken at Brookhaven National Laboratory. The authors acknowledge with pleasure the assistant of Robert Dell and Walter Geisbusch in photographic reproduction, and to Carolyn Trager Burr and Richard Ruffing for specimen preparation. To Helen Lyman we owe thanks for the pencil reconstruction of a cell, and to John Defino for the line drawings of cells in mitosis and for the cover design. We are indebted to Mary Bonneville for careful reading of the text, and to Bonnie Loper and Pamela Pettingill for patient secretarial help. Acknowledgement has been made in the text to those laboratories that have added to the completeness of the book by kindly providing micrographs of special interest. During preparation of the text use was made of the Library of the Marine Biological Laboratory, Woods Hole, for which we express our gratitude. Finally we are indebted to Ralph Wetmore and John Torrey for their valuable criticism of the content.

Myron C. Ledbetter
Keith R. Porter

Table of Contents

Table of Abbreviations and Symbols

BP	Bordered Pit
BP/2	Half Bordered Pit
C	Callose
CB	Cytoplasmic Bridge
CC	Companion Cell
Ch	Chromatin
CI	Cambial Initial
Cp	Chloroplast
CS	Casparian Strip
Ct	Cortex
Cu	Cuticle
CV	Coated Vesicle
CW	Cell Wall
CW_1	Primary Cell Wall
CW_2	Secondary Cell Wall
D	Dictyosome
E	Exine
E_1	Primary Exine
End	Endodermis
ER	Endoplasmic Reticulum
G	Granum
Gr	Granule
GS	Gas Space
I	Intine
L	Lipid
Lo	Lomasome
LP	Lipid-containing Plastid
M	Mitochondrion
Mb	Microbody
ML	Middle Lamella
Mt	Microtubule
N	Nucleus
NC	Nuclear Cap
NE	Nuclear Envelope
Nu	Nucleolus
NuO	Nucleolus Organizer
P	Plastid
Pc	Pericycle
Pd	Plasmodesma
PF	Pit Field
PF/2	Half Pit Field
PGP	Pollen Grain Pore
PM	Plasma Membrane
PP	P-Protein
Pp	Proplastid
R	Ribosome
RC	Ray Cell
RER	Rough Endoplasmic Reticulum
RI	Ray Initial
S	Starch
S_1	First Layer Secondary Wall
S_2	Second Layer Secondary Wall
S_3	Tertiary Wall
SC	Subsidiary Cell
SER	Smooth Endoplasmic Reticulum
SN	Sperm Nucleus
SP	Sieve Plate
ST	Sieve Tube
T	Tonoplast
Ta	Tapetum
To	Torus
Tr	Tracheid
UM	Unit Membrane
V	Vacuole
VN	Vegetative Nucleus
W	Wart
Wx	Wax

Plate 1.1

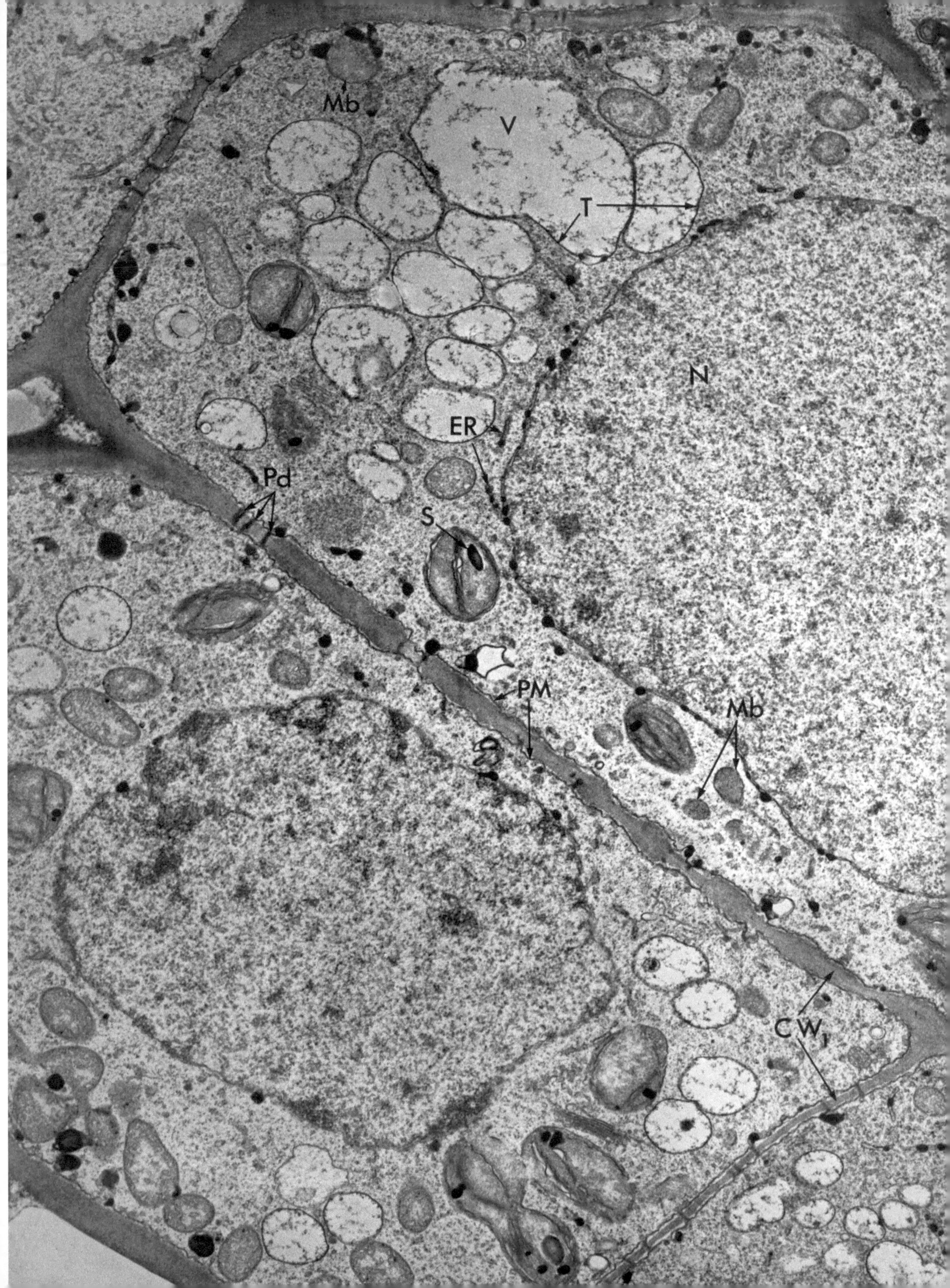

Mb
V
T
N
ER
Pd
S
PM
Mb
CW_1

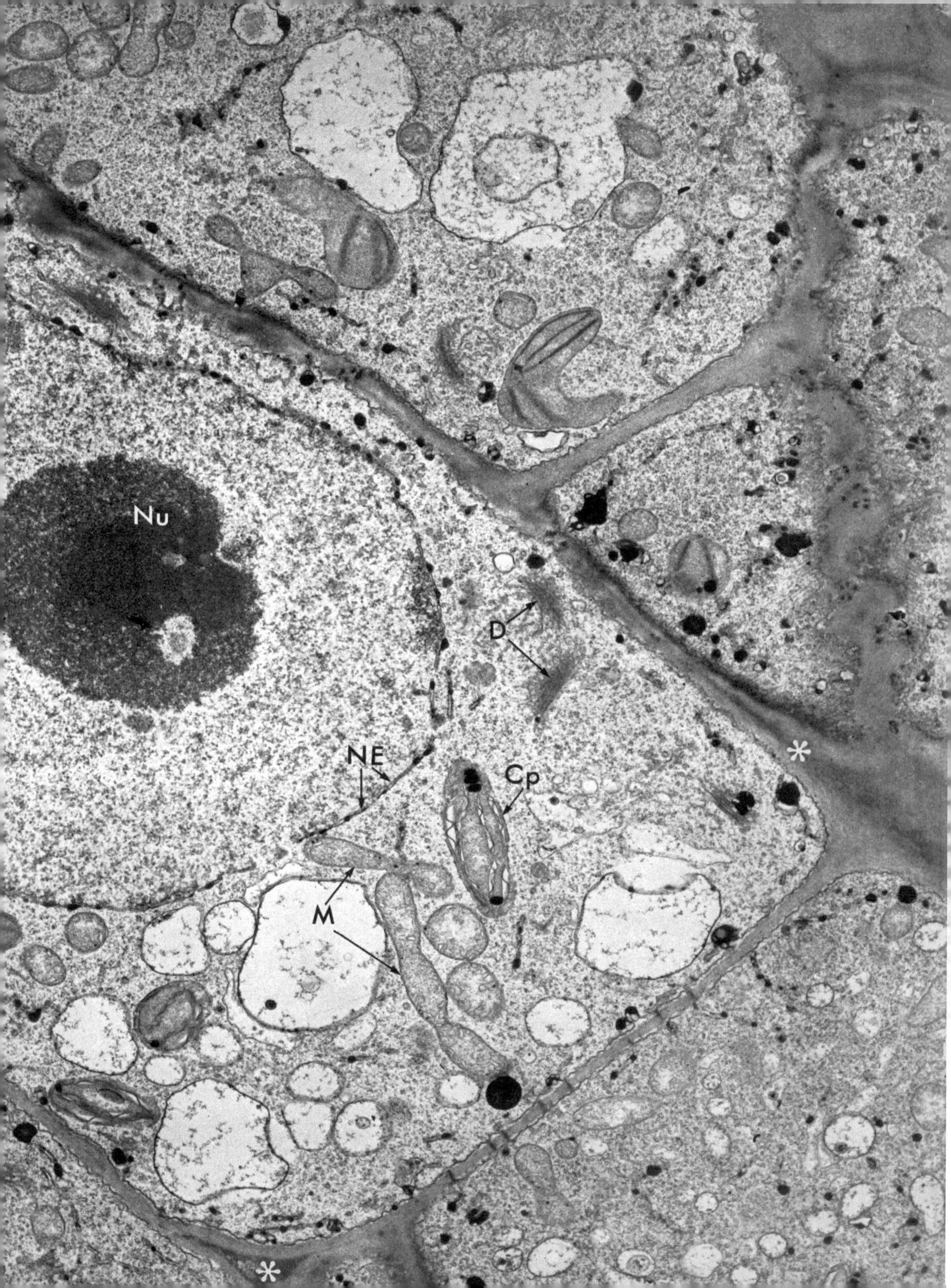

Nu
D
NE
Cp
M
*
*

Plate 1.1

Typical Plant Cell[1]

To the extent that the plant cell can be introduced by a single image, this micrograph serves the purpose. It depicts most of the structures that are commonly found, without, however, showing them in some extreme form characteristic of more highly specialized cells. Subsequent micrographs will illustrate some of the variations that the several parts of a plant cell may adopt in cell differentiation.

The protoplast, as is true of most plant cells, is contained in a cellulose wall (CW) which has been fabricated by the cell itself. It follows that the wall separating two cells is derived in part from each. Frequently these two halves are separated by a third and denser layer, the middle lamella, evident here only at the corners where several cells meet (*).

In this fixed and stained preparation the protoplast is set off from the wall by a thin, dark line which represents the plasmalemma or plasma membrane (PM). This membrane, and closely adjacent materials, constitute the barrier between the living, metabolically active cell and the external environment. It controls what enters and leaves the cell and includes among its several activities the capacity to pump certain ions in and others out against concentration gradients.

The cytoplasm, which of course occupies the zone between the plasmalemma and the nucleus (N), contains several discrete organelles. The mitochondria (M), spherical or cylindrical bodies, 0.5–1.0 micrometers (μm) in diameter, are distributed more or less uniformly throughout the cytoplasm. From counts on a single section and some knowledge of the cell's volume, we can estimate that there are probably 300–400 in a cell of this type. This is in the same order of magnitude as that reported from light microscope studies. As will be seen more clearly in Plate 2.1, each mitochondrion is structurally similar to every other mitochondrion, and these characteristics can be used in the identification of these organelles wherever they occur.

The chloroplasts (Cp), which are without obvious dispositional relation to mitochondria or any other component of the cell, also reside in the cytoplasm and likewise possess a substructure that makes for easy identification. In this micrograph they are relatively small and simple (cf. Plate 8.2), one might say immature, except that in this root tip cell they will never achieve a more mature form. Generally they are larger than mitochondria, have a more uniform size and shape (discoidal) and show internally, parallel arrays of membranes, called lamellae. It is common for chloroplasts to contain starch grains (see Plate 8.1), and in the field of this image at least four are evident (S). In addition the chloroplasts here contain small dense bodies that probably represent lipid (see Plate 8.2).

Prominent also in the cytoplasm are numerous vesicles that appear empty and structureless except for a flocculent precipitate. These represent parts of the vacuole (V), a major component of nearly all plant cells. The separate parts are limited by distinct membranes regarded collectively as part of the tonoplast

[1] Fixed with glutaraldehyde and OsO_4 and stained with uranyl and lead salts.

(T). The origin of this structure is uncertain, but its presence in successive generations of plant cells seems to depend on its being transmitted as an organelle to the daughter cells during cytokinesis.

Small stacks, clusters or bunches of thin lamellar vesicles, scattered about the cytoplasm, make up the dictyosomes (D) (see Plate 2.2). These structures seem to be disposed randomly in this undifferentiated interphase cell.

As part of the complex of organelles, granules and systems that make up the plant protoplast there is an additional element, a population of microbodies (Mb). Here these appear as small spherical units about one half the diameter of a typical mitochondrion. They are limited by a single membrane and show a content that is usually homogeneous but may contain a crystalline inclusion or nucleoid. At the present time there is accumulating evidence that they contain catalase and other enzymes involved in the metabolism of glycolate. They are therefore properly equated with the peroxisomes of animal cells.

The nucleus occupies a central position in this cell, which is to say that it has not been crowded to one side by the expanding vacuole as in most mature plant cells. The nucleus is limited by a double line, which represents an extremely thin (25 nm) lamellar vesicle, the nuclear envelope (NE). Though not clearly evident in this picture because of the low magnification, this envelope (as in animal cells) is perforated at many points by pores (see Plate 2.1).

The sectional image of the nuclear envelope is reproduced at many places in the cytoplasm by what appear to be short segments of the envelope. Actually these are tubular and lamellar extensions of its outer membrane which, considered together, constitute a complex three-dimensional lattice of vesicles and tubes that extends to all parts of the cytoplasm. This system, called the endoplasmic reticulum (ER), is present in all eukaryotic cells. In this cell small dense globules of what is thought to be lipoprotein commonly occur within the vesicles of the envelope and also in the internal phase of the ER. It is easy in this instance to identify parts of the reticulum by these granules. Presumably the lipoprotein is available for incorporation into the forming membranes of the cell during its rapid growth.

The rest of the cytoplasm is populated by small dense particles, the ribosomes, which are shown to better advantage in subsequent micrographs (see Plate 2.1).

The nucleus contains a large nucleolus (Nu), as is common in meristematic plant cells. It has a dense center, and a less dense peripheral zone. The nucleoplasma, apart from the nucleolus, is homogeneous and finely dispersed except for a few scattered blotches of condensed chromatin, called heterochromatin.

The continuity of the end walls of these cells, and to a lesser extent the side walls, is interrupted at numerous points by slender densities which seem to cross the wall (Pd). These are plasmodesmata (see Plate 4.1), which do in fact connect the plasma membranes and cytoplasms of adjacent cells and tend to make of the plant a huge syncitium.

Cell from the root tip of *Elodea canadensis* Michx.

Magnification × 16,000

1.1 Supplemental Reading

Buvat, R.: Electron microscopy of plant protoplasm. Int. Rev. Cytol. **14**, 41–157 (1963).

— Plant cells. New York: McGraw-Hill 1965.

Clowes, F. A. L., Juniper, B. E.: Plant cells. Oxford: Blackwell Scientific Publications 1968.

DeDuve, C., Baudhuin, P.: Peroxisomes (Microbodies and related particles). Physiol. Rev. **46**, 323–357 (1966).

Esau, K.: Plant anatomy, 2nd ed. New York: John Wiley & Sons, Inc. 1965.

Frederick, S. E., Newcomb, E. H.: Cytochemical localization of catalase in leaf microbodies (peroxisomes). J. Cell Biol. **43**, 343–353 (1969).

Frey-Wyssling, A., Mühlethaler, K.: Ultrastructural plant cytology. New York: Elsevier Publishing Co. 1965.

Mollenhauer, H. H., Morre, D. J., Kelley, A. G.: The widespread occurrence of plant cytosomes resembling animal microbodies. Protoplasma (Wien) **62**, 44—52 (1966).

O'Brien, T. P., McCully, M. E.: Plant structure and development. London: The MacMillan Co./Collier & MacMillan, Ltd. 1969.

Voeller, B. R.: The plant cell: Aspects of its form and function. In: The Cell, vol. 6, p. 245–312 (Brachet, J., and Mirsky, A. E., eds.). New York: Academic Press 1964.

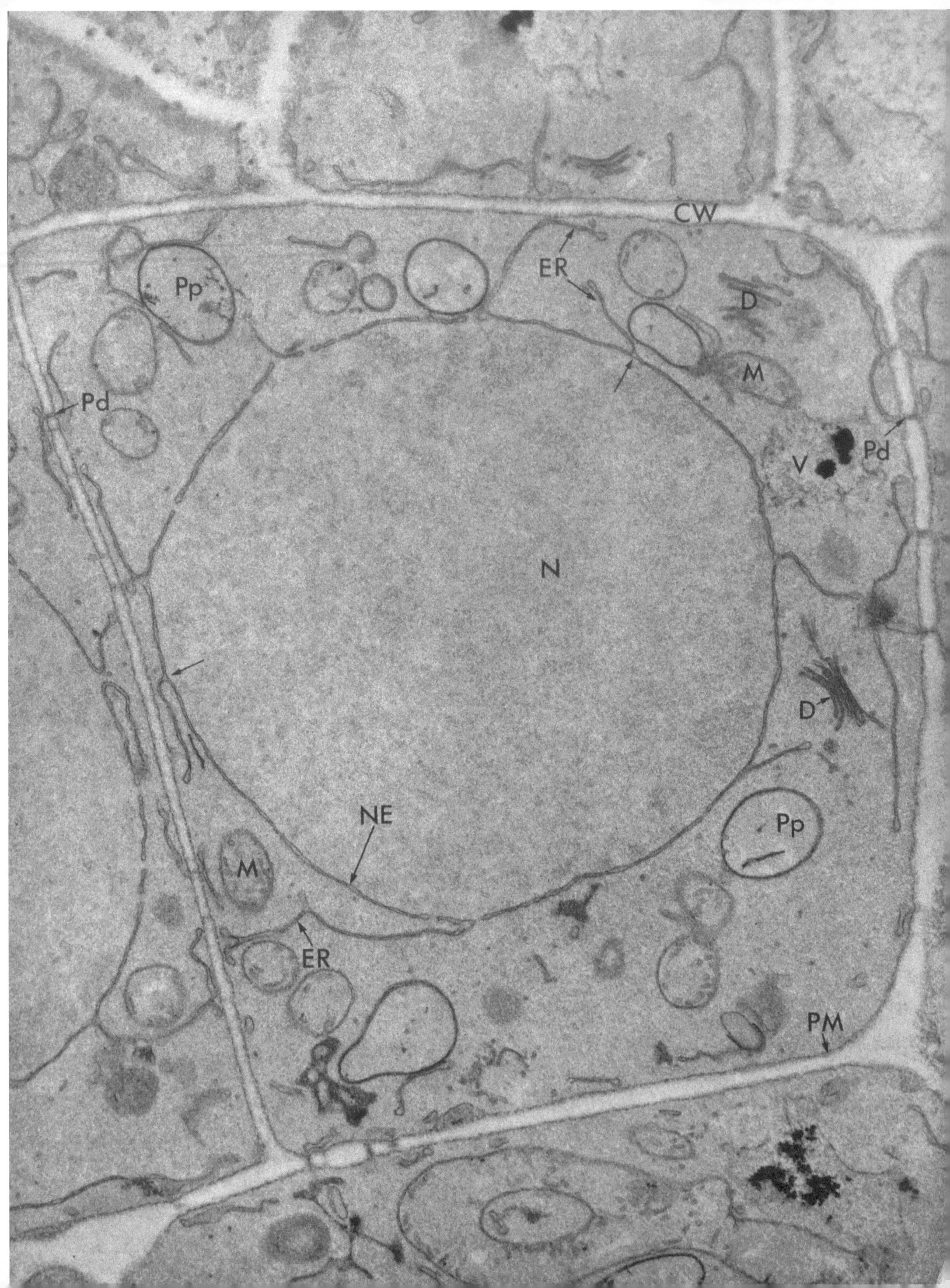
CW
ER
Pp
D
M
Pd
Pd
V
N
D
NE
Pp
M
ER
PM

Plate 1.2

Meristematic Plant Cell, Permanganate Fixed

The electron microscopist is obliged, by the characteristics of his microscope, to study very thin sections of cells previously fixed, dehydrated and embedded in plastic or resin. At best, therefore, the images are not better than a representation of the morphology of the living unit. Obviously it becomes important to reduce artifacts to a minimum, and toward this end one seeks to find and use as fixatives reagents that immobilize the major macromolecular species in their native locations and prevent dislocations and dissolution during subsequent dehydration and embedding. When one considers the complex nature of the system being fixed and the various barriers which the plant cell erects against the penetration of toxic and destructive agents, it is remarkable that any chemical reagents have been found that will produce an adequate fixation.

A representative image of a meristematic cell after fixation in osmium tetroxide is shown in Plate 2.1, and for comparison that of a similar cell after potassium permanganate is shown in this micrograph. There are obviously some striking differences that are important to note, because $KMnO_4$ has been widely employed in plant cell fixation.

Osmium tetroxide was first used in the fixation of animal cells for electron microscopy because, on the basis of early light microscopy, it was known to provide a remarkably faithful preservation. The results for electron microscopy were better than anticipated. Besides achieving a good preservation of form and apparently bonding macromolecules of proteins and lipoproteins into stable gels, the osmium atom with its electron-scattering capabilities was deposited differentially in the cell and acted as a stain by enhancing the apparent density of those structures most reactive with the tetroxide. It was soon found, however, to have its drawbacks. There was no clear endpoint in fixation, and so the quality of preservation varied with time and, in fact, decreased noticeably after extended periods of exposure to the fixative. More serious, possibly, was the slow penetration of OsO_4 that allowed time for cells centrally located in the tissue block to react to the unusual conditions of the environment before being fixed. For plant tissues the rate of penetration was even slower than that for animal material, and this factor made discouraging the early attempts to get good preservation. It was subsequently found that more satisfactory results could be achieved by using phosphate (or cacodylate) instead of veronal buffer as a vehicle, especially in the case of blocks of tissues, in which the cells were surrounded mostly by thin primary walls.

During the early phases of these fixation experiments, potassium permanganate was resurrected from the experience of earlier light microscopists. It was found to penetrate rapidly and to preserve membranous structures selectively. The images were diagrammatic in their appearance and therefore exciting. Successful where osmium had failed, it was applied extensively for a brief period, and much of the early literature reporting electron microscopy of plant cells was based on permanganate-fixed material. It is now known to combine selectively with biological membranes and to stabilize the continuity of these during alcoholic dehydration, when in fact most of the fixation, in the usual sense, is achieved. During exposure to permanganate, however, several structures of the cytoplasmic ground substance and nucleus may disintegrate (depending somewhat on the time the fixative is allowed to act) and even the contours and distribution of membranous structures may change.

Examination of this micrograph, especially after studying the preceding and following micrographs bears out the above remarks on fixation. The nucleus (N) can be identified by its circular profile and central location in this meristematic cell. It lacks evidence of internal structure here (i. e. nucleolus, heterochromatin, etc.), probably because of its relatively long exposure (20 hours) to the fixative. Following a shorter period, some variation in density, representing the distribution of hetero- and euchromatin, is usually evident, but the relative density of these two components reverses during fixation and finally ends in the essentially blank image depicted here.

The nuclear envelope (NE), which marks the limits of the nucleus, clearly consists of two membranes separated by a space that is remarkably uniform in width. Continuities of the envelope with profiles of the cytoplasmic endoplasmic reticulum (ER) are common. In places (arrows) these are seen as outfoldings or evaginations of the outer membrane of the envelope, and such images doubtless approximate the native state. Probably less reliable are instances in which the profiles of both membranes of the envelope extend out into the cytoplasm and appear to extend from one margin of a nuclear pore. These images favor the idea mentioned above that "permanganate-fixed" membranes can "flow" before finally being stabilized by alcohol.

Other membrane-limited structures of the cytoplasm can be identified by those characteristics of their morphology that seem to survive the fixation. Mitochondria (M) come through as spherical or oblate bodies with a few small cristae. Proplastids (Pp) are larger with more prominent membranes and only sparse evidence of lamellae. Both of these organelles display zones of lower density in an otherwise homogeneous matrix, and one may surmise that these coincide with the distribution of DNA-rich compounds. Dictyosomes (D) appear in various views, uniform in showing closely stacked thin vesicles.

The cell wall (CW) exists here as a layer of low density between the opposing plasma membranes of the two cells. When plant cell walls are heavily impregnated with lignin, they appear (upon reaction with the fixative) strikingly dense. In favorable presentations the plasmodesmata (Pd) show as tiny channels between adjacent protoplasts, limited by continuations of the cell membranes.

Although not especially reliable as a fixative, permanganate has the virtue of preserving selectively the membrane-limited components of the cell. Its use makes their identification relatively easy and facilitates studies of organelle populations or relatively gross morphological changes that occur with changes in function or during differentiation. It represents, therefore, even in its limited application, a valuable reagent in the armamentarium of the plant cytologist. Both osmium and permanganate have more recently been superceded as general fixatives by glutaraldehyde, formaldehyde, acrolein or combinations of these with osmium.

From sporogenous tissue of *Saintpaulia ionantha* Wendl.

Magnification $\times 23{,}000$

1.2 Supplemental Reading

Luft, J. H., Permanganate—a new fixative for electron microscopy. J. biophys. biochem. Cytol. **2**, 799–802 (1956).

Porter, K. R., Machado, R. B.: Studies on the endoplasmic reticulum. IV. Its form and distribution during mitosis in cells of onion root tip. J. biophys. biochem. Cytol. **7**, 167–180 (1960).

Whaley, W. G., Mollenhauer, H. H., Leech, J. H.: The ultrastructure of the meristematic cell. Amer. J. Botany **47**, 401–449 (1960).

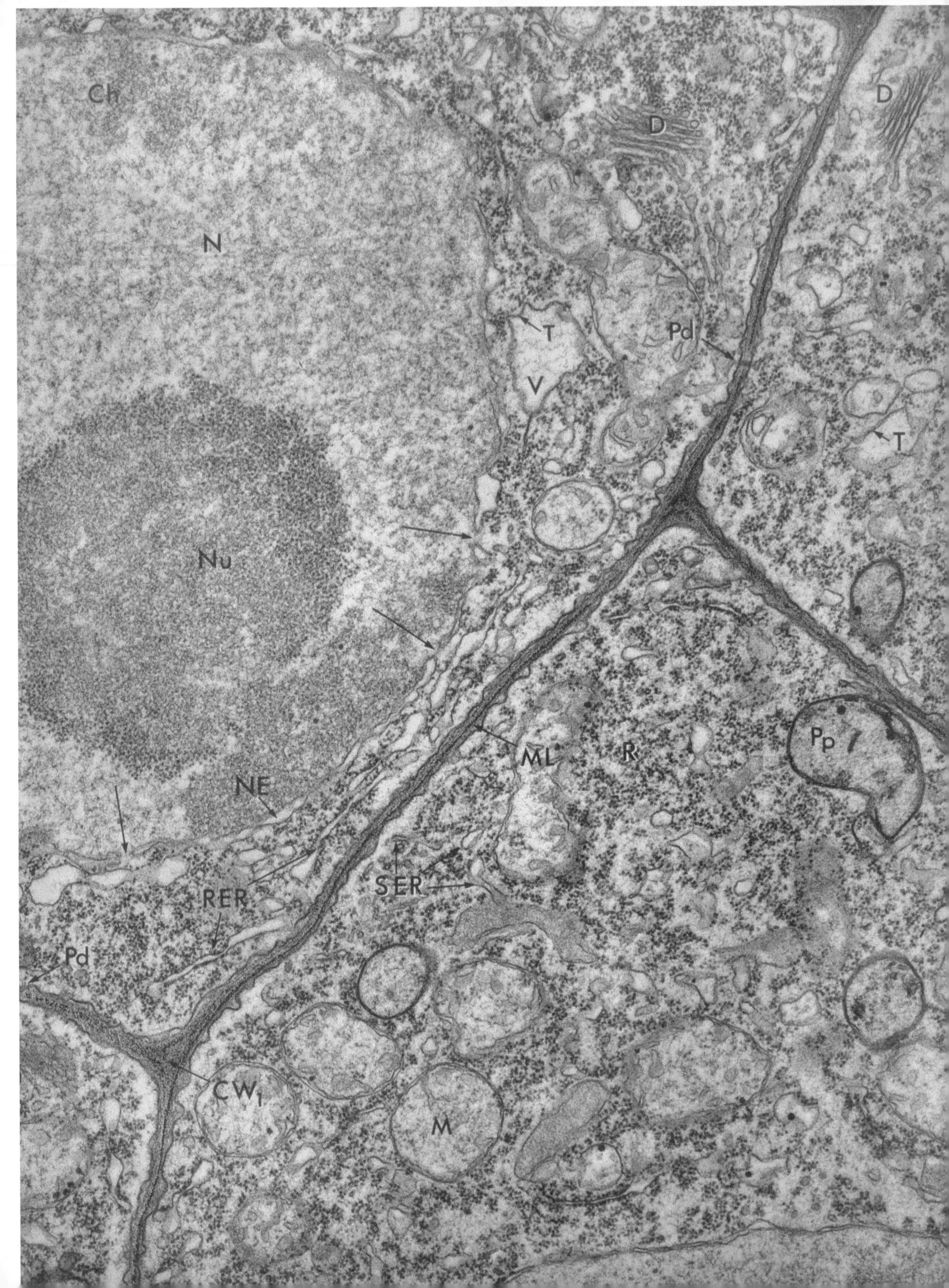

Ch
D
D
N
T
Pd
V
T
Nu
ML
R
Pp
NE
SER
RER
Pd
CW1
M

Plate 2.1

Organelles of the Meristematic Cell

When it became evident that sections of suitable thinness could be prepared for electron microscopy, an attempt was made to extend the exploration of cell fine structure to plant cells and tissues. As already mentioned, the problems of fixation, embedding and sectioning seemed more formidable than in the case of animal tissues, because of the cutinized surfaces and lignified walls which delayed penetration of the fixative and subsequently made difficult the physical act of sectioning. Recourse was taken initially to the study of meristematic cells, where the walls are thinner and the vacuole less prominent. Thereafter, as better fixations and embedding matrices (araldite and epoxy resins) were introduced and as diamond knives became available, the investigation of plant tissues ranged more broadly, and now essentially none of these various botanical materials remains beyond the techniques of electron microscopy.

For the purpose of introducing plant cell fine structure, the meristematic cell is pedagogically ideal. It contains all of the more commonly occurring organelles and protoplasmic systems in a relatively simple and undifferentiated form. It is, of course, typical only of a meristematic (embryonic) plant tissue, which in plant development quickly gives way to one specialization or another.

This micrograph shows parts of four meristematic cells as found in the relatively undifferentiated sporogenous tissue of African violet anthers. The cells are separated by thin primary walls (CW_1). One of the cells (upper left) shows a large nucleus (N) with prominent nucleolus (Nu); the others are represented mostly by regions of cytoplasm containing organelles and systems that are, for the most part, common to both plant and animal cells.

Certainly the most prominent component of the meristematic plant cell is the nucleus. In a cell that measures approximately 8×10 micrometers (μm), the nucleus may be 6 μm in diameter and so constitutes about 15% of the volume of the cell. Such prominence is typical of relatively undifferentiated meristematic cells. Characteristically the genetic material appears in two states of condensation: the relatively dense masses (Ch), representing the presumably inactive chromatin (heterochromatin) and the less dense, seemingly more hydrated or "puffed" chromatin, which is interpreted as being the more active in DNA transcription. The distribution and ratios of condensed to non-condensed chromatin vary from one cell type to another in plants as in animals and have become in both an expression of nuclear differentiation.

The nucleolus is structured around a central core, the nucleolar organizer, which is similar in texture to and continuous with an adjacent marginal condensation of chromatin. Both the central core and the marginal condensation represent segments of a chromonema, the part in the nucleolus being called the nucleolonema. The outer cortical zones of the nucleolus are obviously composed of dense particles, about 150 Å in diameter, which approach in size and density the ribosomes (R) of the cytoplasm. This so-called pars amorpha of the nucleolus is sloughed off when meristematic cells go into prophase, and the ribosome-like particles eventually intermingle with the spindle substance and with the cytoplasm. In telophase the pars amorpha is reconstructed by the synthesis of new granules.

The nucleus is set off from the cytoplasm by a nuclear envelope (NE). This structure, long recognized in the light microscope image of these cells, was not depicted or even imagined in its "true" form until examined by electron microscopy. Here it appears as composed of two membranes separated by an intervening space. Actually this is the sectional view of a large lamellar cistern or vesicle that surrounds the whole nucleus. It is not, however, a complete cover, for in places its continuity is punctuated by fenestrations or pores (see arrows). These seem to exist as channels of continuity between the nucleoplasm on one side and the cytoplasmic ground substance on the other. These pores tend to be of uniform size, about 150 nanometers (nm) in diameter, and evidence of some organization in their structure at the macromolecular level (e.g., an octagonal outline) is accumulating. Moreover, the lumen of the pore is not without content. Sometimes this takes the form of a diaphragm, which has a demonstrable capacity to impede the free passage of large molecules (e.g. ferritin) and colloidal paticles between the interphase nucleus and the surrounding cytoplasm.

The cytoplasm is a conglomeration of vesicles, particles, tubules, mitochondria and proplastids, all of which are suspended in the continuous phase or ground substance. The mitochondria (M) are perhaps the most prominent of these components. They vary from spherical to cylindrical in form and tend toward uniform diameters in the range of 1.0 μm. These ubiquitous cell organelles display the same characteristic structure here as in animal cells. They are limited by two distinct membranes, and from the inner of these small infoldings extend into the contained matrix. These infoldings or cristae have the form of shelves and seem designed to increase the surface area of the inner membrane and thereby provide for the spatial distribution of enzymes involved in oxidative phosphorylation. The number of cristae in mitochondria varies directly with the metabolic activity of the cell in which they are located. Thus in the phloem of *Cucurbita*, the companion cells and parenchymal cells show mitochondria with many cristae, whereas in the immediately adjacent and metabolically quiescent sieve tubes the mitochondria are essentially devoid of cristae.

The degree of autonomy of mitochondria in their growth and differentiation is not entirely understood. It is now clear, however, that they possess their own package of DNA, which apparently codes for structural proteins in the organelle and possibly a few of their enzymes. The fluffy material in the matrix of the mitochondria represents in part nucleoproteins. It is also evident from work on *Neurospora* that mitochondria increase in size by the incorporation of materials from the cytoplasmic pool and multiply in number by simple fission. Though there may be instances when this pattern varies, the origin of mitochondria or plastids from other sources, such as the nuclear envelope, has not been established. Other components of mitochondria, e.g., the small, dense ribosome-like particles in the matrix that are known to contain RNA, remind one that these organelles have all the machinery for independent synthesis and reproduction. This fact calls to mind the early speculation that mitochondria first entered into their present state as symbionts, i.e., as distinct and separate organisms initially resembling, perhaps, present day bacteria.

The similarities between mitochondria and plastids have not gone unnoticed. In this micrograph the latter are small and relatively undifferentiated compared with the mature and functional form they are capable of achieving (see Plate 8.2). Thus in this meristematic cell, we can refer to them as proplastids (Pp) without knowledge of whether these in fact could have developed into functioning chloroplasts. Like mitochondria they have two limiting membranes, which are slightly thicker and denser (in the stained preparation) and closer together than those of the mitochondria. Similarly also the inner of the two of the limiting membranes extends into the stroma of the chloroplast (see Plate 2.3). This is the earliest expression of the complex membranous system that eventually makes up the membranous stacks or grana, of the mature chloroplast. Even in these immature and essentially quiescent organelles, there may appear various storage granules of starch (see Plate 1.1) or lipoprotein (see Plates 1.1 and 2.3).

Among the few features that characterize plant cells as distinct from animal cells, one of the most prominent is the vacuole (V). This extraordinary component of the plant protoplast frequently occupies 70–90 percent of the cell's volume and by its expansion compresses the nucleus and the residual, metabolically active cytoplasm into a thin layer against the cell wall. In younger cells, such as those depicted here, the vacuole is probably reticular and may indeed be made up of several discontinuous elements. Thus in a thin section one can identify several profiles of vacuoles (V). Identification is more difficult than in the case of mitochondria because vacuoles have no consistent internal structure. There are, however, one or two helpful features. The membrane, for example, is generally more prominent than those of other cytoplasmic elements, and in the young cell this membrane, called the "tonoplast" (T), appears too large for the vacuole contents at this time. Thus it peaks or folds outward, frequently giving to the vacuole profile a star shape. The tonoplast seems to have many interesting properties. It is probably self-propagating if, of course, the required building blocks are available. It has some very distinctive permeability properties, which define it as selectively permeable and capable of active transport. Thus certain waste products of metabolism are stored in the vacuole and transported across the tonoplast against gradients of concentration. Seemingly the vacuole serves as a sink for all manner of compounds from calcium oxalate to anthocyanins and tannins (Plates 9.2 and 9.3), and the tonoplast determines what shall enter the sink and what shall be retained there.

The origin of the vacuole is a topic of great interest to plant cytologists. Unfortunately good observations are still in short supply. Possibly it develops as a special differentiation of the endoplasmic reticulum, homologous to the so-called smooth ER of the animal cell. On the other hand, there is equally good evidence indicating its existence as an independent organelle, a system of the plant cell which is passed on from cell to cell in small fragments and which subsequently, under the conditions prevailing in that cell, expands to specialize in this or that function.

Other membrane-limited spaces within the cytoplasm are readily identified by the experienced microscopist as part of the endoplasmic reticulum. As already mentioned (see Plates 1.1 and 1.2), this is a system of vesicles and channels that are at times demonstrably continuous with one another and with the nuclear envelope. The limiting membrane is thin (50–60 Å) compared with that surrounding the protoplast (80–100 Å). The vesicles are frequently flattened, and in this state often develop fenestrae at their margins to give a reticular arrangement of channels. It is common for these vesicles, particularly the larger flattened cisternae, to be covered on their cytoplasmic surfaces by dense particles, the ribosomes. This combination, known here as in animal cells as rough-surfaced endoplasmic reticulum (RER), appears to be designed primarily for the synthesis of proteins. The product of synthesis is initially sequestered in the cavities of the system and from these it is either selectively retained for intracellular use or is secreted to the extracellular environment. The population of ribosome-free vesicles (SER) appears to be derived from the rough endoplasmic reticulum. Some of these especially near the protoplast surface, may be important in providing enzymes for the initial phases of cellulose synthesis.

The biogenesis and assembly of the ER membranes, which serve as surfaces for the disposition of ribosomes, is only now being investigated. It is known that with each division of the cell, fragments of the ER are passed intact to each daughter cell. There they reassemble to form a nuclear envelope and a new labyrinth of cytoplasmic channels. But, for the growth of membrane that follows, the source of building blocks is in doubt. Conceivably they could either derive from the synthetic activity of encrusting ribosomes and be stored as lipoproteins in the cavities of the ER or, just as possibly, they could come from the free ribosomes that populate the cytoplasmic ground substance.

The ribosomes (R) are among the most remarkable of all macromolecules. They represent a precise assembly of ribonucleic acid and perhaps 20 different proteins. When linked to information-bearing RNA (messenger), they effect the assembly of long chains of amino acids into proteins. It is thought that the amino acid sequences are achieved on the ribosome and that the secondary and tertiary foldings of the polypeptide chains occur in the cavities of the ER cisternae or in the cytoplasmic ground substance. In this micrograph we can identify also small clusters of ribosomes, called polysomes. Although no micrograph could demonstrate it, one assumes on the basis of other findings that many of these clusters are identical not only in the arrangement of ribosomes but also in their association with messenger RNA (see Plate 2.4). Such a group of polysomes, it is postulated, whether attached to membranes or free, is involved in making the same protein or series of proteins.

This description of the plant cell and the endoplasmic reticulum has not been exhaustive as far as the membrane-rich systems of the cell are concerned. Apparently Nature discovered early the value of membrane surfaces for the distribution of macromolecules involved in synthesis and transport, to mention only two biological processes associated with membranes. The multibranched, polymorphic form as well as the wide cytoplasmic distribution of the ER brings its surfaces within short diffusion distances of metabolites in the cytoplasmic ground substance and of those entering the cell. But equally well adapted to the requirement for being where needed, or where metabolites are, are the many mitochondria and plastids.

Units of the Golgi complex, also called dictyosomes (D), constitute a system that is actively involved in the synthesis of wall materials and extracellular matrices for the plant protoplast. In this micrograph the individual dictyosome appears as a stack of unconnected flat vesicles, each about 250 Å thick. The number of vesicles per dictyosome varies but is frequently around 6 or 7 as illustrated here. The number of dictyosomes also varies, tending to be larger in cells most actively engaged in wall production. These and other features of the Golgi complex are enlarged upon in Plate 2.2 and the associated legend.

To be complete this discussion of the cytoplasm should include several other components. Lysosomes, for example, while not common in cells of the sporogenous tissue, are commonly found in other plant cells. They are rich in phosphatases and other hydrolytic enzymes and appear to perform the same functions of focal and controlled autolysis as their counterparts in animal cells. Microtubules, seemingly engaged in form-molding activities, are not evident here because osmium tetroxide, used as a fixative, does not preserve them in plant cells (see Plate 2.4).

A distinct line, the plasma membrane, limits each cell in the micrograph. It appears to be continuous, uniform in thickness and of wavy contour. Its origins require more investigation but evidence available indicates that small vesicles originating in the Golgi are fed into this surface.

The cells of this sporogenous tissue are separated by extremely thin (65 nm) primary walls (CW_1). In osmium-fixed material it is difficult to discern any significant texture in the wall, except that provided by the pectinaceous material making up the dense middle lamella. Plasmodesmata (Pd) appear in a few places. From sporogenous tissue of *Saintpaulia ionantha* Wendl.

Magnification ×34,000

2.1 Supplemental Reading

Hsiao, T. C.: Characteristics of ribosomes isolated from roots of *Zea mays*. Biochim. biophys. Acta (Amst.) **91**, 598–605 (1964).

Hyde, B. B.: Changes in nucleolar ultrastructure associated with differentiation in the root tip. J. Ultrastruct. Res. **18**, 25–54 (1967).

Lafontaine, J. G.: Structure and mode of formation of the nucleolus in meristematic cells of *Vicia faba* and *Allium cepa*. J. biophys. biochem. Cytol. **4**, 777–784 (1958).

— Chouinnard, L. A.: A correlated light and electron microscope study of the nucleolar material during mitosis in *Vicia faba*. J. Cell Biol. **17**, 167–201 (1963).

Luck, D. J. L.: The influence of precursor pool size on mitochondrial composition in *Neurospora crassa*. J. Cell Biol. **24**, 445–460 (1965).

— Formation of mitochondria in *Neurospora crassa*. A study based on mitochondrial density changes. J. Cell Biol. **24**, 461–470 (1965).

Luck, D. J. L., Reich, E.: DNA in mitochondria of *Neurospora crassa*. Proc. nat. Acad. Sci. (Wash.) **52**, 931–938 (1964).

Nadakuvkaren, M. J.: Fine structure of negatively stained plant mitochondria. J. Cell Biol. **23**, 193–195 (1964).

Newcomb, E. H.: Fine structure of protein-storing plastids in bean root tips. J. Cell Biol. **33**, 143–163 (1967).

Palade, G. E.: Intracisternal granules in the exocrine cells of the pancreas. J. Cell Biol. **2**, 417–422 (1956).

Porter, K. R.: The ground substance; observations from electron microscopy. In: The cell, vol. 2, p. 621–676 (Brachet, J., and Mirsky, A. E., eds.). New York: Academic Press 1961.

Sun, C. N.: Fine structure of the root cells of *Phaseolus vulgaris*. I. Structure of the meristematic cells. Cytologia (Tokyo) **27**, 204–211 (1962).

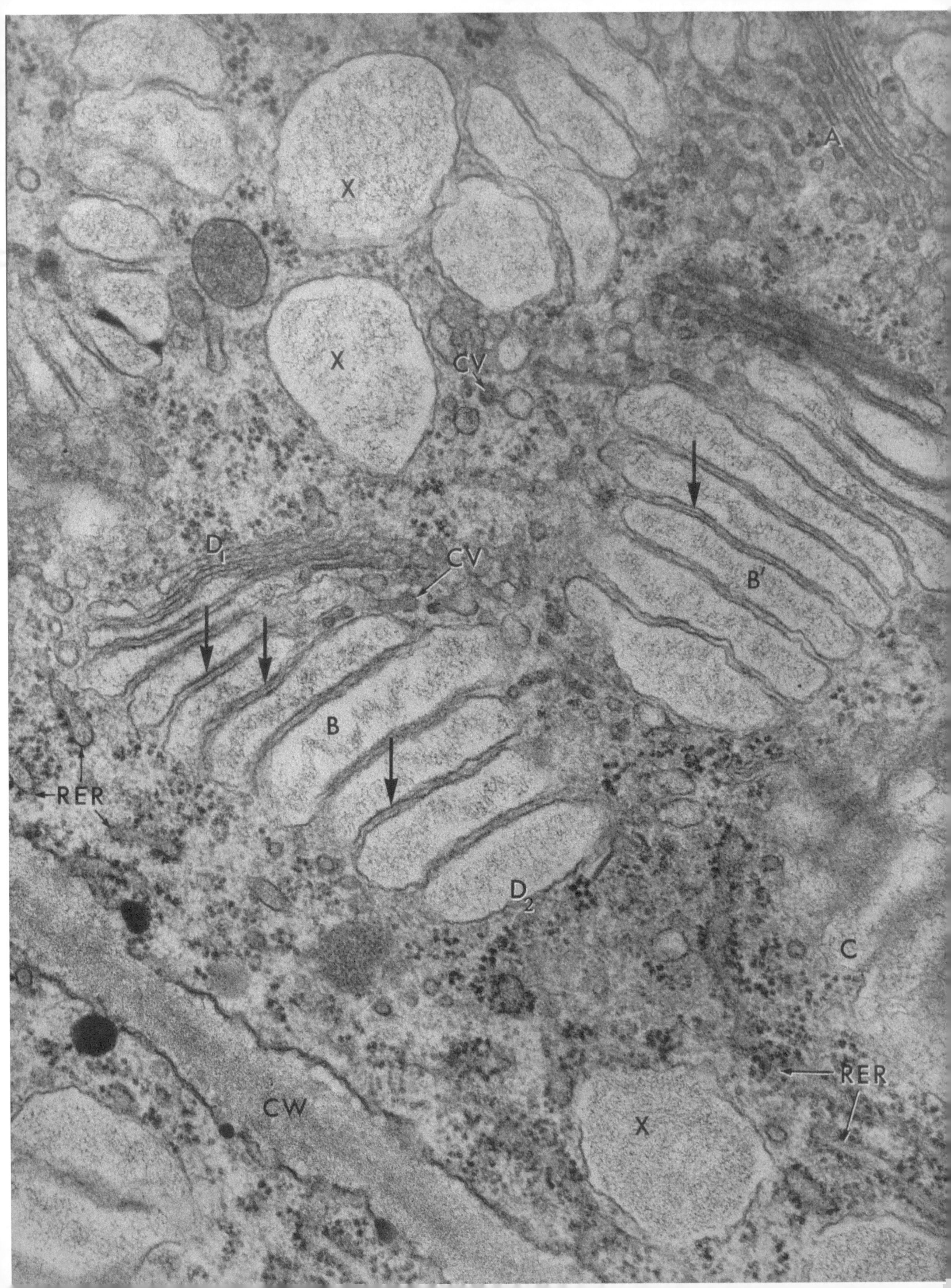

A
X
X
CV
D1
CV
B′
B
RER
D2
C
RER
CW
X

Plate 2.2

Dictyosomes

One of the major activities of plant cells is the production of the cellulose capsule, which encloses the cell and contributes strength and rigidity to plant stems. The precursors of this wall are assembled from glucose in the plant protoplast and discharged into the environment where other wall ingredients and enzymes cooperate in the formation and organization of the slender cellulose filaments that characterize all plant cell walls. The deposition of precursor materials and the formation of walls can be extraordinarily rapid, as the elongation of root hairs and the growth of pollen tubes attest. It has been calculated that a single tracheid cell of xylem may deposit micrograms of cellulose in a 24-hour period. This kind of activity naturally depends on the integrated functioning of several intracellular systems and organelles. Among these the dictyosome is recognized as being intimately involved in the final acts of assembly and packaging-for-export of precursor wall materials. Vesicles, loaded with polysaccharides of various kinds, form from the Golgi cisternae and move either to the cell surfaces where the wall is being assembled or to the surfaces of slime-producing cells of the root cap, depicted in this micrograph. Such cells as these, because of their hyperactivity in producing cell coatings rich in carbohydrates, are outstanding objects for studying the form and function of the dictyosome.

The typical plant dictyosome appears in vertical section as a stack of thin flat vesicles, resembling wafers (D in Plate 2.1). There are at least 6 or 7 of these vesicular units in each complex. The spaces between them are also thin and fairly uniform (about 120 Å). The individual units or cisternae are, however, not identical. Indeed, within a stack they display a gradient of forms from one pole (to be called the forming face) to the other (the maturing face). Those at the forming face usually have a smaller diameter than those at the opposite face. The limiting membranes of the cisternae increase in prominence (thickness) from top to bottom of the stack, i.e., from the forming to the maturing face.

The individual Golgi cisterna is characteristically thin and of uniform thickness in its central, medullary zone (Plate 2.1). Its surface there is continuous and apparently as unadorned structurally as most smooth biological membranes. Toward the margins of this wafer-like structure, however, holes or fenestrae appear, which give this region a reticulate form as shown in text Fig. 2.2a. At the outer limits of this marginal zone, at least along part of their perimeter, a few cisternae have associated with them a population of small thick-walled vesicles, which in most instances appear coated on their outer surfaces. These coated vesicles (CV) are about 85 nm in diameter including the coat. They have been observed in various degrees of continuity with the margins of the larger cisternae and are interpreted as forming at this site from these elements (text Fig. 2.2a).

Around this general form, the dictyosome displays several variations, and one is characteristic of the slime-secreting cell of the root cap shown here. In this micrograph the cell surface is represented at the lower left as it faces a primary cell wall (CW). The cytoplasm above and to the right includes several profiles of the endoplasmic reticulum with associated ribosomes (RER). The micrograph is dominated, however, by dictyosomes, comprising stacks of vesicles. At A there is a complex (or stack) of relatively typical (in the sense defined above) cisternae with associated small vesicles, whereas at B and B′, where the plane of section is somewhat marginal to the central axis of the stack, only one or two of the cisternae have retained their thin wafer-like form with fenestrated margins. These are at the forming face (D_1) of the stack. The others, numbering nine, have vesiculated over the entire diameter of the individual cisterna and contain an obvious accumulation of what is assumed to be slime material. They can be seen to grow progressively thicker toward the maturing face of the stack (D_2). It is indeed from such images as these that the impression has developed that in the living state new cisternae are constantly forming at one pole, from ER-derived vesicles (D_1), and are being sloughed off as secretory vesicles from the opposite pole (D_2). Smaller coated vesicles (CV) appear at several places within the micrograph and in some instances are still associated with other elements of the dictyosome.

Some of the cisternae (X), apparently inflated with secretory products, have adopted a spherical form, have separated from the stack and are headed for the cell surface to discharge their contents. Others, also substantially filled, but still associated with a stack (as at D_2) are separated from each other by a relatively uniform distance of about 25 nm. This space is in some planes of sectioning clearly and precisely bissected by a thin line of moderate density (arrows). Where similar cisternae have been cut obliquely (as at C) instead of vertically, this intercisternal layer is seen to be finely filamentous, with the filaments arranged roughly parallel to each other and evenly spread within the thin blanket. It has been postulated, on the basis of the spacings involved, that this blanket-like layer represents the protein of an extra lipoprotein layer that is organized between the cisternae. One could with equal conviction propose that it represents a compressed layer of the cytoplasmic matrix or ground substance, with which it is continuous, and that the fabric-like structure is representative of a similar three dimensional lattice that exists elsewhere in the ground substance but is ordinarily difficult to visualize in the depth dimension of a thin section. Regardless of the interpretation, this intercisternal layer is apparently common to all images of plant dictyosomes when fixed with glutaraldehyde.

Among the several roles that have been assigned to intracytoplasmic structures, none is more firmly established than the involvement of the Golgi complex in the synthesis of extracellular polysaccharides, or more specifically in this instance, the pectins. Autoradiography of root tips after exposure to tritiated D-glucose has clearly demonstrated this fact. Subsequent isolation and study of the radioactive polysaccharide has shown that the D-glucose is incorporated into the glucosyl residues of the polymer cellulose. There is uncomplicated evidence that radioactivity is concentrated first in the dictyosomes and subsequently in the cell walls and secreted slime. In animal cells, a similar involvement of the Golgi in the synthesis of polysaccharide-rich cell coats and intercellular ground substances (hyaluronic and glucuronic acid) permits the generalization that the Golgi complex plays principally this role in the cell's economy, with involvement as well in such related functions as the assembly of plasma membranes having special functions.

From the cap of a young growing root of *Phleum pratense* L.

Magnification × 80,000

2.2 Supplemental Reading

Bisalputra, T., Ashton, F. M., Weier, T. E.: Role of dictyosomes in wall formation during cell division of *Chlorella vulgaris*. Amer. J. Botany **53**, 213–216 (1966).

Bonnett, H. T., Newcomb, E. H.: Coated vesicles and other cytoplasmic components of growing root hairs of radish. Protoplasma (Wien) **62**, 59–75 (1966).

Bonneville, M. A., Voeller, B. R.: A new cytoplasmic component of plant cells. J. Cell Biol. **18**, 703–708 (1963).

Grove, S. N., Bracker, C. E., Morré, D. J.: Cytomembrane differentiation in the endoplasmic reticulum-Golgi-apparatus-vesicle complex. Science **161**, 171–173 (1968).

Manton, I.: On a reticular derivative from Golgi bodies in the meristem of *Anthoceros*. J. biophys. biochem. Cytol. **8**, 221–231 (1960).

— Leedale, G. F.: Observations on the fine structure of *Paraphysomonas vestita*, with special reference to the Golgi apparatus and the origin of scales. Phycologia **1**, 37–57 (1961).

Mollenhauer, H. H.: An intercisternal structure in the Golgi apparatus. J. Cell Biol. **24**, 504–511 (1965).

Mollenhauer, H. H., Morré, D. J.: Tubular connections between dictyosomes and forming secretory vesicles in plant Golgi apparatus. J. Cell Biol. **29**, 373–376 (1966).

— Whaley, W. G.: An observation on the functioning of the Golgi apparatus. J. Cell Biol. **17**, 222–225 (1963).

Morré, D. J., Mollenhauer, H. H.: Isolation of the Golgi apparatus from plant cells. J. Cell Biol. **23**, 295–305 (1964).

Newcomb, E. H.: A spiny vesicle in slime-producing cells of the bean root. J. Cell Biol. **35**, C 17–C 22 (1967).

Northcote, D. H., Pickett-Heaps, J. D.: A function of the Golgi apparatus in polysaccharide synthesis and transport in the rootcap cells of wheat. Biochem. J. **98**, 159–167 (1966).

Pickett-Heaps, J. D.: Further observations on the Golgi apparatus and its functions in cells of the wheat seedling. J. Ultrastruct. Res. **18**, 287–303 (1967).

Turner, F. R., Whaley, W. G.: Intercisternal elements of the Golgi apparatus. Science **147**, 1303–1304 (1965).

Whaley, W. G., Kephart, J. E., Mollenhauer, H. H.: Developmental changes in the Golgi-apparatus of maize root cells. Amer. J. Botany **46**, 743–751 (1959).

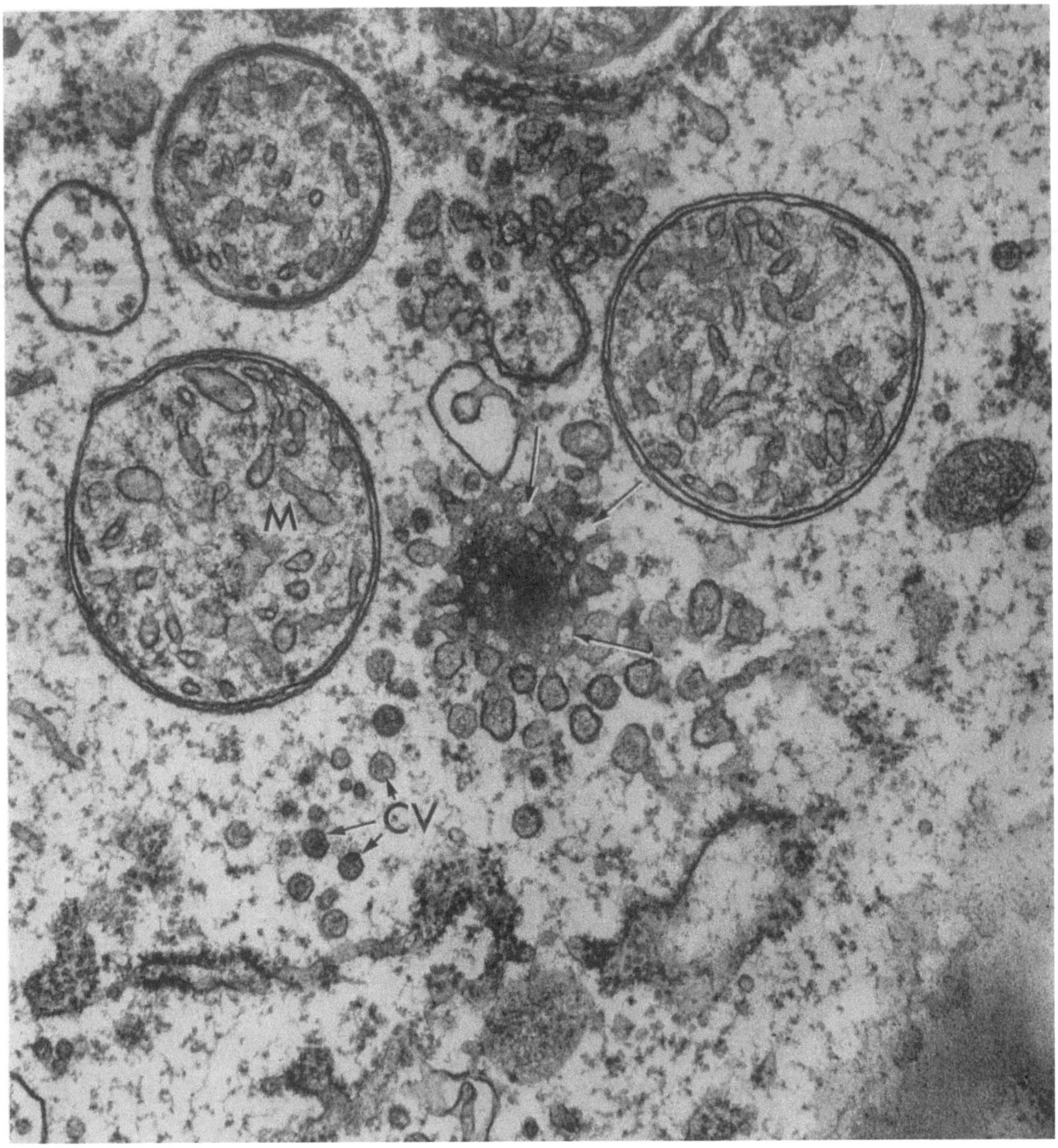
M
CV

Text Fig. 2.2a

Dictyosome in Face View

This micrograph provides a "dorsal" view of a dictyosome and helps to clarify the form of the individual cisterna. Here two or three of these are caught within a single section, and even though the image is of several superimposed, the individual cisterna emerges as being solid in the center, fenestrated toward the margin (arrows) and transformed into separate thick-walled vesicles at the outside (CV). This same structure has been observed in meristematic cells of *Anthoceros* and in Golgi complexes isolated intact from maize root tips.

From mesophyll cells of leaf of wheat, *Triticum aestivum* L.

Magnification $\times$ 50,000

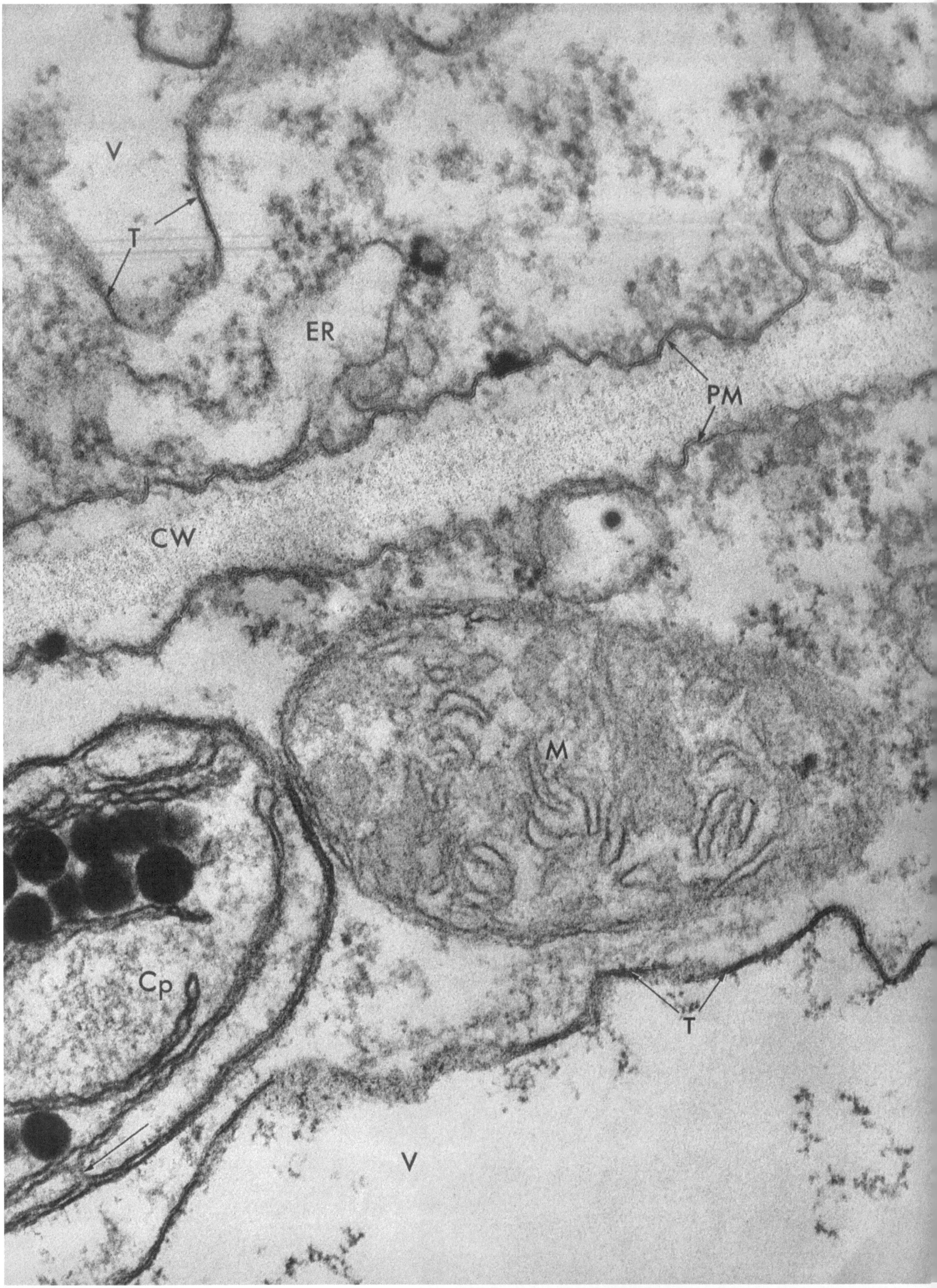
V
T
ER
PM
CW
M
Cp
T
V

Plate 2.3

Membranes of Cell and Cell Organelles

The movement of metabolites, salts and water in and out of cells is influenced by a thin layer of phospholipid and protein, the plasma membrane (PM), which marks the boundary of the living protoplast. Some ions and molecular species move with remarkable ease through this barrier either by diffusion or along osmotic gradients; others are taken up selectively and transported even against a concentration gradient; and still others are blocked completely. There is ample reason, therefore, to regard this membrane as more than a bimolecular layer of phospholipids and proteins that are uniformly distributed in the plane of the membrane. Instead it apparently has built into it sites for active transport with enzymes for catalyzing the events associated with transport. Pinocytosis as a mechanism for uptake in intact plant cells is not as yet clearly established.

Though one can speak with some certainty about the functioning of the plasmalemma of plant and animal protoplasts, there is less information available on other membranes of the plant cell. That they are all morphologically similar to the PM and constructed on the same basic 3-layered plan can be observed from any of these micrographs, and on this basis they might be casually judged to possess other properties not dissimilar to those which characterize the plasmalemma. This is, however, specious reasoning. It is much more obvious that the intracellular membranes provide the protoplast with large areas of surface and limit compartments that can serve as reaction chambers and storage depots. Though possessing in common a trilaminar structure, these membranes are found on close examination to differ in dimensions (thickness), and this feature serves to set them apart morphologically from one another.

Shown here at relatively high magnifications are small parts of two adjacent parenchymal cells from a wheat flower filament. A portion of the cell wall (CW) separating the two protoplasts crosses the micrograph just above the center. It is bordered on each side by the plasma membranes of the cells (PM). Where these membranes are viewed directly on edge they appear to consist of two dense lines with an intervening space and thus to possess the so-called unit membrane structure. Their overall thickness is approximately 80 Å. Of the two lines the outer one seems to be slightly the more prominent. In this construction, the dense lines are interpreted as representing osmium-fixed protein in fusion with the hydrophilic ends of phospholipids. The intervening space (20–25 Å thick) is occupied, according to current dogma, by the long chain fatty acids of the phospholipids. There is occasional evidence of beading along the dense lines, but whether this deserves any attention in a preparation so harshly treated before microscopy and then exposed to the damaging action of the electron beam seems doubtful.

With this morphology and these reservations in mind one can usefully compare the plasma membrane with others in the cell. For example, that limiting the vacuole (V), and known as the tonoplast (T), is slightly thicker (100 Å) in these cells than the plasma membranes. The same trilaminar structure is evident, but the innermost line (that facing the cavity of the vacuole) is much the more prominent. The membranes in the mitochondria (M), although apparently uniform within that organelle, are strikingly thinner (60 Å) than both types just described. Again the trilaminar structure is evident as in the other membranes, but it is more difficult to resolve. The membrane limiting the vesicles and channels of the endoplasmic reticulum closely resembles the mitochondrial type. Chloroplasts (Cp), of which one is shown in part at the lower left, are different in having membranes that are intermediate in thickness (75 Å) between the plasma membrane and those of the ER and in having an intervening lamina of low density so thin as to be almost unresolvable. The lamellar membranes within the proplastid are identical with those limiting it, a fact that is not surprising since the inner of the two and the membrane of the lamellae are continuous (arrow).

As mentioned above these morphological differences probably reflect functional differences, which may in fact be very striking. One has to keep in mind that distinctions made on the basis of thickness and relative prominence, though seemingly slight, are doubtless major at the molecular level, describing as they do a possible twofold difference in size of certain of the membrane subunits.

From stamen filament of wheat, *Triticum aestivum* L.

Magnification × 130,000

2.3 Supplemental Reading

Branton, D.: Membrane structure. Ann. Rev. Plant Physiol. **20**, 209–238 (1969).

Higginbotham, N., Hope, A. B., Findlay, G. P.: Electrical resistance of cell membranes of *Avena* coleoptiles. Science **143**, 1448–1449 (1964).

Korn, E. D.: Structure of biological membranes. Science **153**, 1491–1498 (1966).

Ledbetter, M. C.: Observations on membranes in plant cells fixed with OsO_4. In: Proc. 5th Intern. Congr. Electron Microscopy, vol. 2, W-10 (Breese, S. S., Jr., ed.). New York: Academic Press 1962.

Maddy, A. H., Malcolm, B. R.: Protein conformations in the plasma membrane. Science **150**, 1616–1618 (1965).

Marchant, R., Robards, A. W.: Membrane systems associated with the plasmalemma of plant cells. Ann. Botany (Lond.) **32**, 457–471 (1968).

Mayo, M. A., Cocking, E. C.: Pinocytic uptake of polystyrene latex particles by isolated tomato fruit protoplasts. Protoplasma (Wien) **68**, 223–230 (1969).

Northcote, D. H. (ed.): Structure and function of membranes. Brit. med. Bull. **24** (2), 99–184 (1968).

Sjöstrand, F. S.: A new ultrastructural element of the membranes in mitochondria and of some cytoplasmic membranes. J. Ultrastruct. Res. **9**, 340–361 (1963).

Staehelin, L. A.: Ultrastructural changes of the plasmalemma and the cell wall during the life cycle of *Cyanidium caldarium*. Proc. roy. Soc. B **171**, 249–259 (1968).

Stoeckenius, W., Engelman, D. M.: Current models for the structure of biological membranes. J. Cell Biol. **42**, 613–646 (1969).

Yamamoto, T.: On the thickness of the unit membrane. J. Cell Biol. **17**, 413–422 (1963).

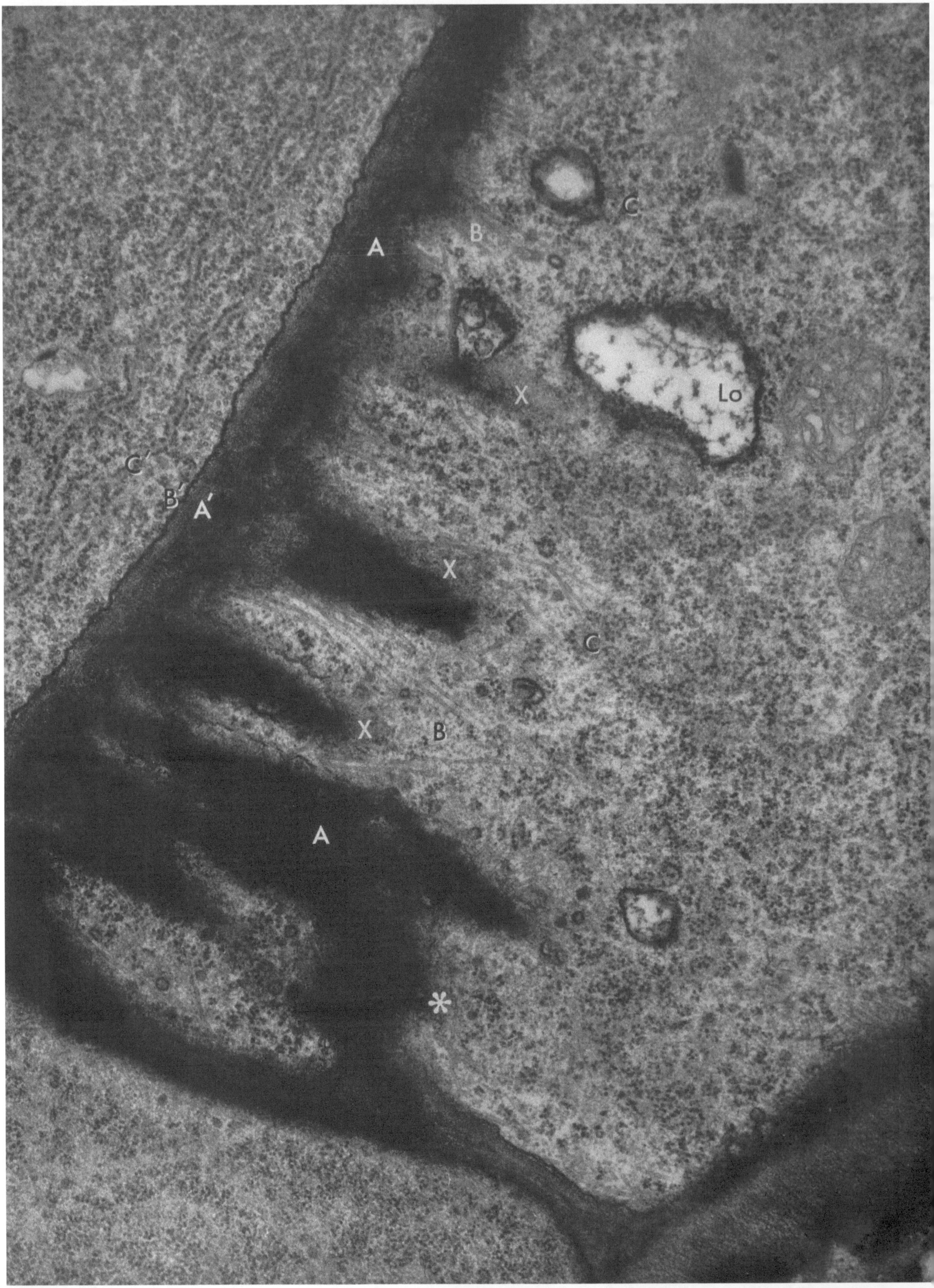
A
B
C
X
Lo
C′
B′
A′
X
C
X
B
A
*

Plate 2.4

Microtubules in the Cell Cortex

The discovery of slender microtubules as ubiquitous structures in plant protoplasts was made only recently and came as something of a surprise. There had been no indication of their existence from light microscopy, and the early electron micrographs of cells fixed with either osmium or permanganate had shown nothing of the sort. It appears now that, behaving as extraordinarily labile structures, the microtubules had simply disassembled in the presence of these other "fixatives".

The individual microtubule or filament is an unusual structure in several respects. It is uniform in its diameter (about 240 Å) over its entire length, which may be as much as several micrometers. For a structure so slender it is remarkably straight and seems, indeed, to exist in many instances as an elastic rod. It never branches. In its size, form and other characteristics (see below) it closely resembles the "filaments" in the "9+2" complex of the cilium and flagellum. Despite minor differences in their lability, structures with the characteristics of microtubules are conceived of as homologous. They are now known to occur widely in eukaryotic cells and are essentially as general as mitochondria and ribosomes.

In this micrograph of a meristematic tissue the end of one cell appears in a grazing section (tangential to a side wall), which includes part of the wall (A), an expanse of the cell cortex (B) and a larger expanse of cytoplasm just within the cortex (C). Above and to the left an adjacent cell is caught in more nearly medial longitudinal section, and the zones that are spread out widely in the central image are here contained within a zone not thicker than 1 μm (at A′, B′ and C′). The dense material under A and A′ represents the stained cell wall, within which a substructure of fine fibrils is sometimes evident. Between it and zone B or B′, we can identify the plasma membrane as a dense line limiting the cell at the upper left or as a gray area in the grazing section (under X's). Zones B and B′ are revealed as thin layers of lower density, thickly populated by microtubules. The latter tend to run parallel to one another just beneath the plasma membrane. As oriented here they run around the cylindrical protoplast in planes roughly perpendicular to the long axis of the cell. Near the cross-wall end of this cell, included at the lower left, the orientation of microtubules becomes more random (as at *).

In the cell at the upper left, where the section transects vertically the cytoplasmic cortex, the microtubules are shown in cross section (B′). They are obviously present here in large numbers (ca. 20 in a one micrometer length of cortex) and actually in numbers greater than evident here because they show clearly in an electron micrograph only if viewed directly on end, i.e., oriented parallel to the electron beam. It is also apparent that in this case they are confined to the thin cortical zone (ectoplasm) of this plant protoplast and run circumferentially around the cell.

The C zone of the central cell includes surface views of lamellar cisternae of the endoplasmic reticulum, similar to those cisternae that are caught in vertical section in the cytoplasm at the upper left (C′). Ribosomes associated with membranes of the ER are arranged in repeating configurations (spirals, whorls, rosettes), which are interpreted as polysomes (see Plate 2.1). The prominent patches of low density (Lo) containing a flocculent material and looking like vacuoles actually represent sections through shallow invaginations of the cell surface. Such miniature lakes external to the plasma membrane have been called lomosomes and may be sites of wall formation.

The distribution of microtubules in plant cells is frequently similar to that pictured here, but substantial and doubtless significant variations from this have been observed. Firstly this arrangement is most characteristic of meristematic cells in interphase. During division phases, the tubules dramatically vanish from the cell cortex and appear in the mitotic spindle as the spindle fibers. Even before mitosis begins in some cells, the microtubules shift their distribution to form a recognizable band in a position that anticipates the subsequent location of the cell plate and plane of division. During later stages of differentiation, the microtubules appear specifically in bands or bundles and in locations and orientations that coincide with the secondary thickenings in the cell wall. They would seem from this association to be related in some way to the process of wall formation. In long slender trichomes, the tubules may be arranged in a cylinder with smaller filamentous units as a central core, a possible axial skeleton influencing the shape of the elongating cell process.

The function of microtubules in the plant cell is not obvious. In both plant and animal cells they occur in zones of the cytoplasm that are demonstrably gelled. It seems indeed that they coincide with and possibly constitute a frame for many cytoplasmic gels and that by virtue of their dimensions and elasticity they can give such gels the assymetric forms they possess. This association of microtubules brings them into positions which border on regions of solated cytoplasm, where streaming activity (cyclosis) is most evident. It is not surprising, therefore, that early speculation on the role of microtubules assigned them an active function in streaming. Whether they do in fact provide a motive force (via undulation) or simply a semirigid surface, a skeletal frame, where energy is released to some translational mechanism (a mechanochemical phenomenon) that moves the adjacent solated cytoplasm is still in question. That they do define channels along which the cytoplasm may stream selectively is well established. There is, for example, their interesting positioning over the zones of secondary wall deposition where the cytoplasm actively streams. And it is possible that, like the thick filaments of the myofibril, the surfaces of microtubules possess arms that may through some configurational change favor the motion of the adjacent solated cytoplasm. But just as reasonable, in view of the several known instances (including cilia) where microtubules are part of undulating structures, is the suggestion that streaming is referable to migrating waves of ordering in the material surrounding the microtubules.

The various characteristics of microtubules, as well as their patterned distribution and orientation (usually parallel to the long axes of asymmetric cells) has favored the suggestion that microtubules are also involved in shaping cells. The idea is that they are distributed according to information contained in an as yet undefined system of the cytoplasm. The latter, through morphological and functional design, determines the shape, etc., of the cage constructed of microtubules, and this cage (cytoskeleton) in turn determines the form of the developing plant protoplast after cytokinesis, at a time when the wall is relatively pliant.
From the root tip of *Phleum pratense* L.
Magnification ×39,000

2.4 Supplemental Reading

Cronshaw, J.: Cytoplasmic fine structure and cell wall development in differentiating xylem elements. In: Cellular ultrastructure of woody plants, p. 99–124 (Côté, W. A., Jr., ed.). Syracuse, N.Y.: Syracuse University Press 1965.

— Bouck, G. B.: The fine structure of differentiating xylem elements. J. Cell Biol. **24**, 415–431 (1965).

Dawes, C. J., Barsilott, D. C.: Cytoplasmic organization and rhythmic streaming in growing blades of *Caulerpa prolifera*. Amer. J. Botany **56**, 8–15 (1969).

Hepler, P. K., Newcomb, E. H.: Microtubules and fibrils in the cytoplasm of *Coleus* cells undergoing secondary wall deposition. J. Cell Biol. **20**, 529–533 (1964).

Ledbetter, M. C.: Fine structure of the cytoplasm in relation to the plant cell wall. J. Agr. Food Chem. **13**, 405–407 (1965).

— Porter, K. R.: A "microtubule" in plant cell fine structure. J. Cell Biol. **19**, 239–250 (1963).

Newcomb, E. H.: Plant microtubules. Ann. Rev. Plant Physiol. **20**, 253–288 (1969).

— Bonnett, H. T., Jr.: Cytoplasmic microtubules and wall microfibril orientation in root hairs of radish. J. Cell Biol. **27**, 575–589 (1965).

O'Brien, T. P.: Cytoplasmic microtubules in the leaf glands of *Phaseolus vulgaris*. J. Cell Sci. **2**, 557–562 (1967).

Pickett-Heaps, J. D., Northcote, D. H.: Organization of microtubules and endoplasmic reticulum during mitosis and cytokinesis in wheat meristems. J. Cell Sci. **1**, 109–120 (1966).

Pickett-Heaps, J. D., Northcote, D. H.: Relationship of cellular organelles to the formation and development of the plant cell wall. J. exp. Botany **17**, 20–26 (1966).

Porter, K. R.: Cytoplasmic microtubules and their functions. In: Ciba Foundation Symposium on Principles of Biomolecular Organization, p. 308–345 (Wohlstenholme, G. E. W., O'Connor, M., eds.). London: J.&A. Churchill 1966.

— Ledbetter, M. C., Badenhausen, S.: The microtubule in cell fine structure as a constant accompaniment of cytoplasmic movement. Proc. 3rd European Reg. Conf. Electron Microscopy, vol. B, p. 119 (M. Titlbach, ed.). Prague: Publishing House of the Czechoslovak Academy of Sciences 1964.

Preston, R. D., Goodman, R. N.: Structural aspects of cellulose microfibril biosynthesis. J. roy. micr. Soc. **88**, 513–527 (1968).

Sabnis, D. D., Jacobs, W. P.: Cytoplasmic streaming and microtubules in the coencytic marine alga *Caulerpa prolifera*. J. Cell Sci. **2**, 465–472 (1967).

Shelanski, M. L., Taylor, E. W.: Isolation of a protein subunit from microtubules. J. Cell Biol. **34**, 549–554 (1967).

Tilney, L. G., Gibbins, J. R.: Differential effects of antimitotic agents on the stability and behavior of cytoplasmic and ciliary microtubules. Protoplasma (Wien) **65**, 167–179 (1968).

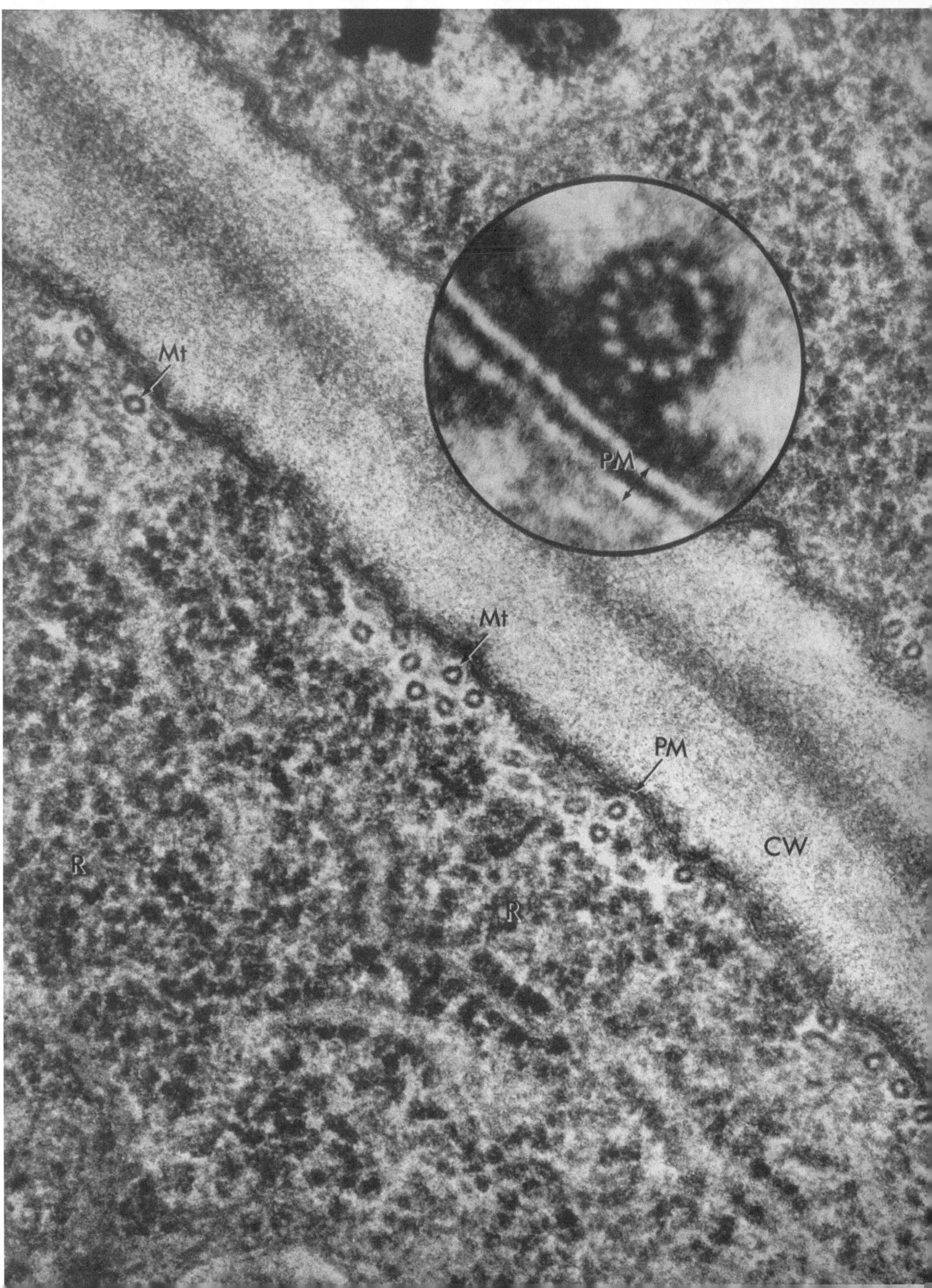

Mt
PM
Mt
PM
CW
R
R

Plate 2.4.1

Molecular Structure of Microtubules

The typical microtubule (Mt) of the plant cell is shown in this micrograph in cross section, where it is part of a population in the cell cortex. A plasma membrane (PM) and cell wall (CW) limit the cortex on one side. Each microtubule is approximately 240 Å in diameter and consists of a wall, about 80 Å thick, and a center of low density 80 Å across. It is evident that each is surrounded by a zone of low density from which such particles as ribosomes (R) are excluded. This indicates that a small region of the adjacent cytoplasmic matrix is organized relative to the surface of the microtubule. Its presence and its thickness (ca. 100 Å) may determine the minimal center-to-center distance between microtubules in this material. Such a separation is not, however, a constant feature in arrays of microtubules. In some examples, found in other cell types, they are attached by structural arms or bridges and are evenly spaced. In other instances, as in the peripheral pairs of cilia, they are so intimately associated that the adjacent surfaces integrate to form a single wall. Thus we recognize that the surfaces of microtubules may vary widely in form and probably in related function.

To what extent these differences in surface characteristics reflect differences in the molecular structure and composition of the microtubules is difficult to assess. It is known, as shown in the insert, that the wall of each microtubule is made up of 13 subunits. These appear in negative image in this particular micrograph because of a dense osmiophilic material that has impregnated the surrounding matrix and to some extent the interior of the tubule. The evenly spaced spots of low density represent the protein molecules which assemble to form the walls of the tubules. They are about 50 Å in cross sectional diameter. Since they disassemble at low temperatures and under high hydrostatic pressures, it is assumed that when assembled into tubules they are hydrogen bonded at hydrophobic sites on their surfaces.

When isolated and viewed in negatively stained preparations, the microtubule shows further details of its substructure. It is clear, for example, that the 13 units observed in cross section represent subfilaments, which appear as rows of uniform beads of macromolecules (spaced at 40 Å) arranged parallel to the long axis of the tubule. Therefore the basic subunit in the construction of the tubule wall measures about 40 by 50 Å. It is known to be a protein and is calculated to have a molecular weight of 60,000. When assembled to form the wall of the microbutule these particles are not in perfect register in planes normal to the axis of the tubule. Actually the transverse rows are about 10° off the normal, an indication that the particles are arranged in a shallow helix. Such an arrangement is consistent with a growth in length by the addition of individual molecules in a spiral sequence at the free end of the tubule.

Microtubules, whether as the axial components of cilia or of spindle fibers in the mitotic apparatus or as neurotubules in animal nerve cells or as components of the plant cell cortex, seem to have the same size and substructure. Variations in their sensitivity to fixatives, to cold, and to pressure, as well as substantive differences in the amount and character of tubule-associated materials and structures, have however been demonstrated. Despite these variations the core structure, the 13-unit wall, seems constant.

From[1] the root tip of timothy, *Phleum pratense* L.
Magnification × 240,000
Insert is from root tip cell of *Juniperus chinensis* L.
Magnification × 1,000,000

[1] Reprinted from Agricultural and Food Chemistry **13**, 405–407 (1965).

2.4.1 Supplemental Reading

Behnke, O.: Evidence for four classes of microtubules in individual cells. J. Cell Sci. **2**, 169–192 (1967).

Burton, P. R.: Substructure of certain cytoplasmic microtubules: An electron microscopic study. Science **154** 903–905 (1966).

Gall, J. G.: Microtubule fine structure. J. Cell Biol. **31**, 639–643 (1966).

Grimstone, A. V., Klug, A.: Observations on the substructure of flagellar fibres. J. Cell Sci. **1**, 351–362 (1966).

Kiefer, B., Sakai, H., Solari, A. J., Mazia, D.: The molecular unit of the microtubules of the mitotic apparatus. J. molec. Biol. **20**, 75–79 (1966).

Ledbetter, M. C., Porter, K. R.: Morphology of microtubules of plant cells. Science **144**, 872–874 (1964).

Marantz, R., Ventilla, M., Shelanski, M.: Vinblastine-induced precipitation of microbutule protein. Science **165**, 498–499 (1969).

Ringo, D. L.: The arrangement of subunits in flagellar fibers. J. Ultrastruct. Res. **17**, 266–277 (1967).

Rosenbaum, J. L., Moulder, J. E., Ringo, D. L.: Flagellar elongation and shortening in *Chlamydomonas*. J. Cell Biol. **41**, 600–619 (1969).

Stephens, R. E.: Reassociation of microtubule protein. J. molec. Biol. **33**, 517–519 (1968).

Wolfe, S. L.: Isolated microtubules. J. Cell Biol. **25**, Pt. **1**, 408–413 (1965).

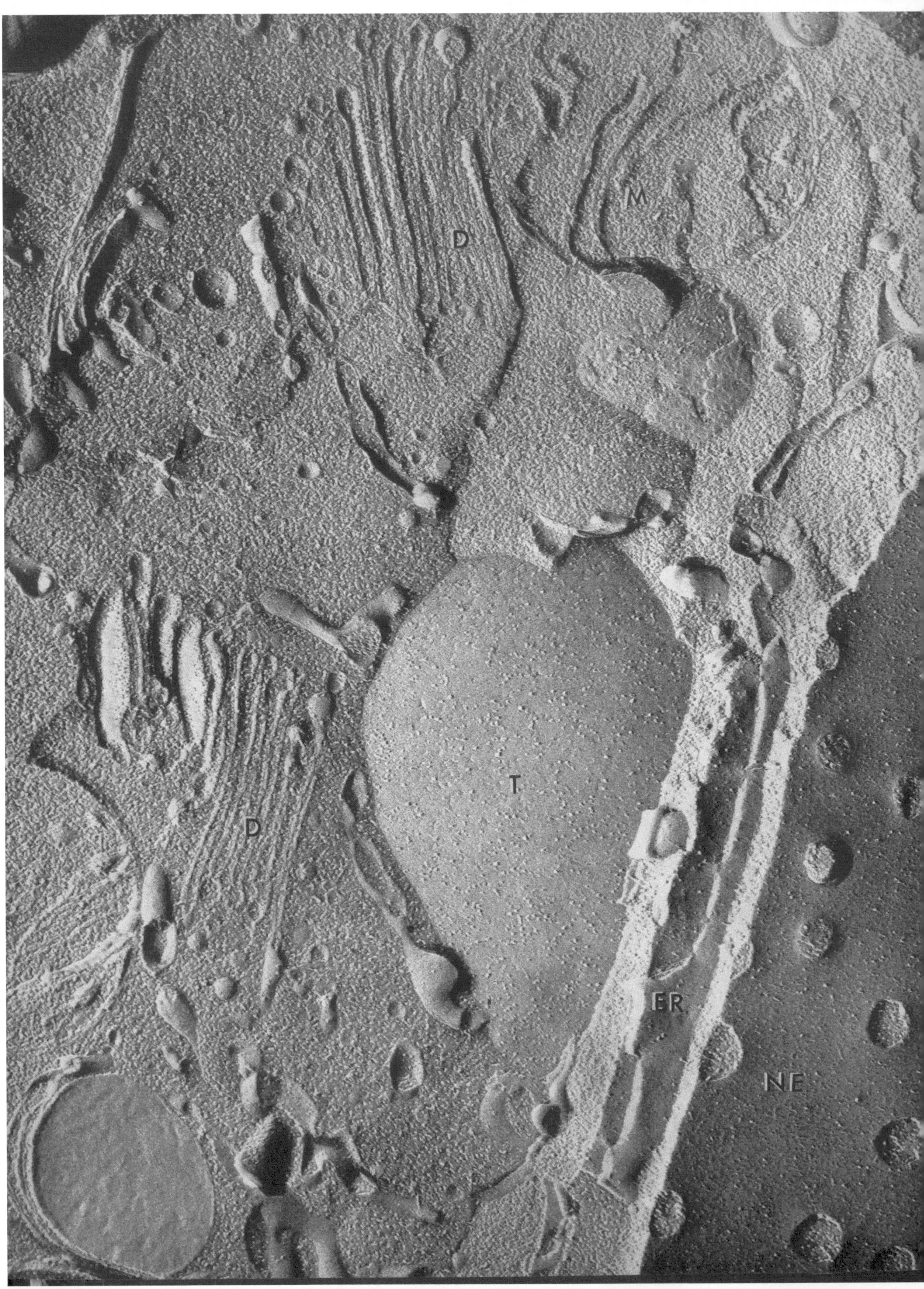
M
D
T
D
ER
NE

Plate 2.5

Freeze-Etch Image of a Cell

The early electron microscope images of plant and animal cells could be related quite satisfactorily to the light microscope images of both fixed and living cells, and so the observer had some confidence that the chemical fixation and plastic embedding essential for electron microscopy had achieved a reasonably faithful preservation of the native form. But some doubts remained, especially in relation to structural elements that could not be resolved by light microscopy. Could the form and disposition of these in the E. M. image be accepted as approximating the same in the living cell? A number of fairly convincing arguments were developed in support of an affirmative answer to this question, and in its grosser aspects, at any rate, the E. M. image was accepted as a useful reflection of the native state. More recently such arguments were rendered largely rhetorical by the development of techniques for studying the fractured surfaces of frozen tissue. Not only did such preparations confirm the existence of the submicroscopic particles, vesicles, filaments and membranes of the conventional E. M. image, but they provided as well excellent surface views of membranes and even a picture believed to represent the material in the layer between the leaflets of the unit membrane.

The freeze-etch technique includes a number of important steps. The tissue to be studied may first have some of its water replaced by glycerol. It is then quickly frozen at liquid nitrogen temperatures in Freon 22. The frozen block is next fractured with a cold knife, and water is encouraged to sublime from the frozen surface (etching). Subsequent to this the surface is shadowed and replicated with platinum and carbon. Observations are made on the replica.

In the micrograph shown here one can recognize with ease the common components of cell fine structure as found in fixed preparations. The lower right includes a part of a nuclear envelope (NE). This is an inside view of the outer membrane. Certain crater-like holes which punctuate the otherwise smooth surface represent nuclear pores. In some instances the content of the pore has fractured external to the level of the membrane, whereas in other instances it has fractured at a level above or flush with the membrane. The particulate and finely fibrous texture of the membrane surface is difficult to interpret but probably reflects some aspects of the membrane's molecular structure.

The large, slightly convex surface (T) more centrally located in the picture is taken to be the outer surface of a vacuole. The population of small particulates on its surface differs from that on the nuclear envelope and probably represents components of the membrane itself exposed by the separation of the inner and outer leaflets of the membrane in the fracture process.

The fracture plane in other parts of the cell has transected membranes to give an edge-on rather than a surface view. This is especially true in the dictyosomes (D). All of the features that from chemically fixed preparations have come to characterize these remarkable complexes are evident as well in the replica of the frozen-etched specimen. Even the trilaminar structure of the unit membranes is resolved. The details of mitochondrial structure (M) are equally well portrayed. An extraordinarily smooth surface is included at the lower left and seems to represent the fracture through a frozen peroxisome. The fact that it is without associated particles makes less likely the interpretation that the particles on other surfaces, as has been claimed, are contaminants of the technique.

Other images of vesicles and surfaces included in the micrograph belong for the most part to the endoplasmic reticulum (ER). Some of these should appear studded with ribosomes but, for some curious reason, possibly their level of hydration, ribosomes fail to show with expected clarity.

The balance of the field represents the cytoplasmic ground substance. It appears uniformly textured with small particulates and equivalent depressions or dimples. These presumably reflect the macromolecular elements of this continuous phase—elements which may or may not be linked into a continuous gel. It is interesting and probably meaningful that the textures of the cytoplasmic and mitochondrial matrices are similar.

From tissues of the onion root tip, *Alium cepa* L. Magnification × 94,000

2.5 Supplemental Reading

Branton, D.: Fracture faces of frozen membranes. Proc. nat. Acad. Sci. (Wash.) **55**, 1048–1056 (1966).

— Moor, H.: Fine structure in freeze-etched *Allium cepa* L. root tips. J. Ultrastruct. Res. **11**, 401–411 (1964).

Moor, H., Muhlethaler, K.: Fine structure in frozen-etched yeast cells. J. Cell Biol. **17**, 609–628 (1963).

Northcote, D. H.: Structure and function of plant-cell membranes. Brit. med. Bull. **24**, 107–112 (1968).

— Lewis, D. R.: Freeze-etched surfaces of membranes and organelles in the cells of pea root tips. J. Cell Sci. **3**, 199–206 (1968).

Staehelin, L. A.: The interpretation of freeze-etched artificial and biological membranes. J. Ultrastruct. Res. **22**, 326–347 (1968).

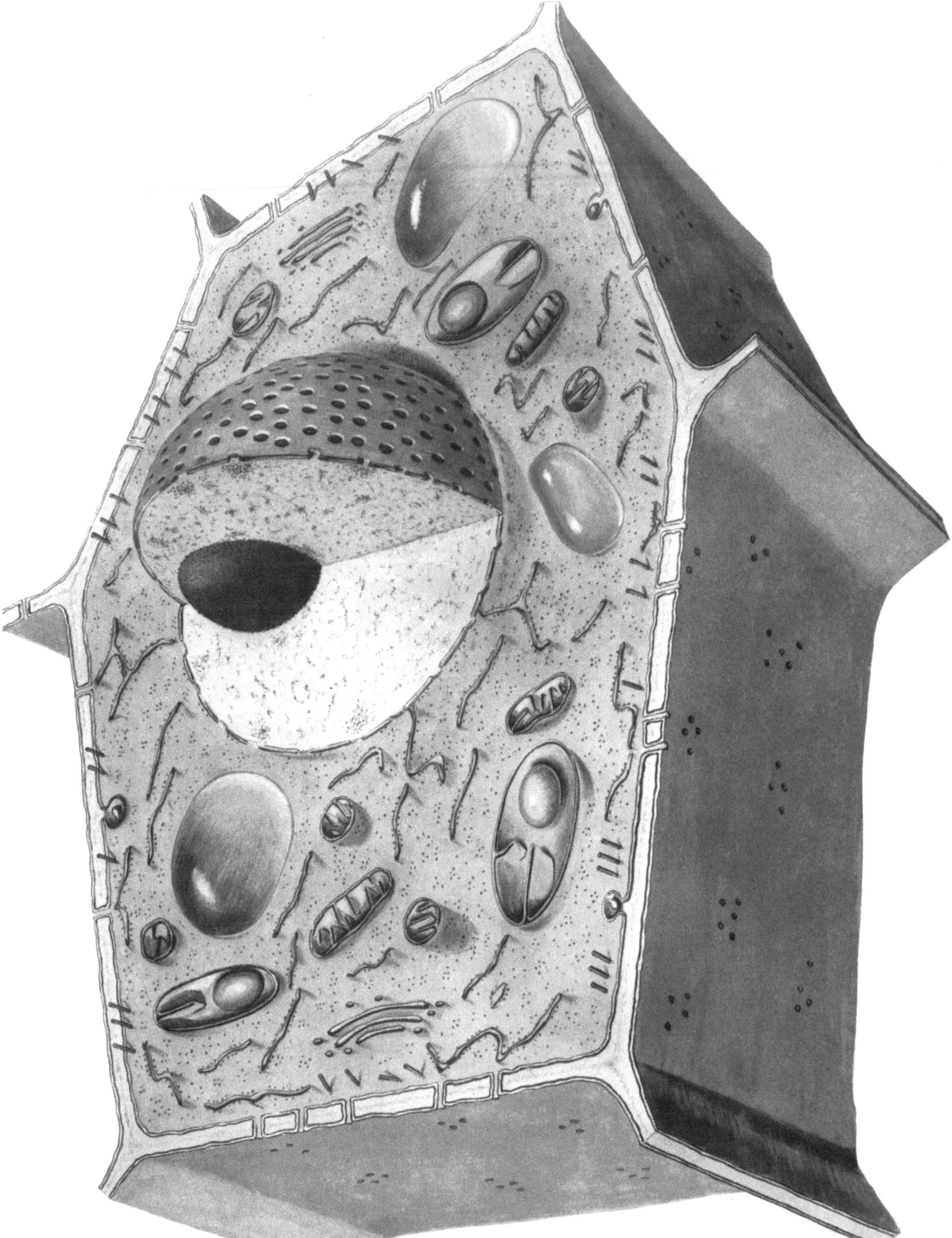

Diagram 2.6

Diagram Including Fine Structure of Typical Plant Cell

Following a study of two dimensional electron micrographs such as that represented by Plates 1.1 to 2.5 it is useful to attempt a three dimensional reconstruction that includes the generalizations made. Such a project becomes an exercise in testing one's understanding and therefore one's ability to communicate the same to an artist, in this instance Helen C. Lyman. This drawing represents then a synopsis of our communication. For the reader of the preceding plates the diagram needs no labels and so none are included. For others their absence may serve as a stimulus to look more carefully into the preceding plates and legends to discover where in fact the authors found the evidence for the generalizations made.

Magnification ca. $\times$ 10,000

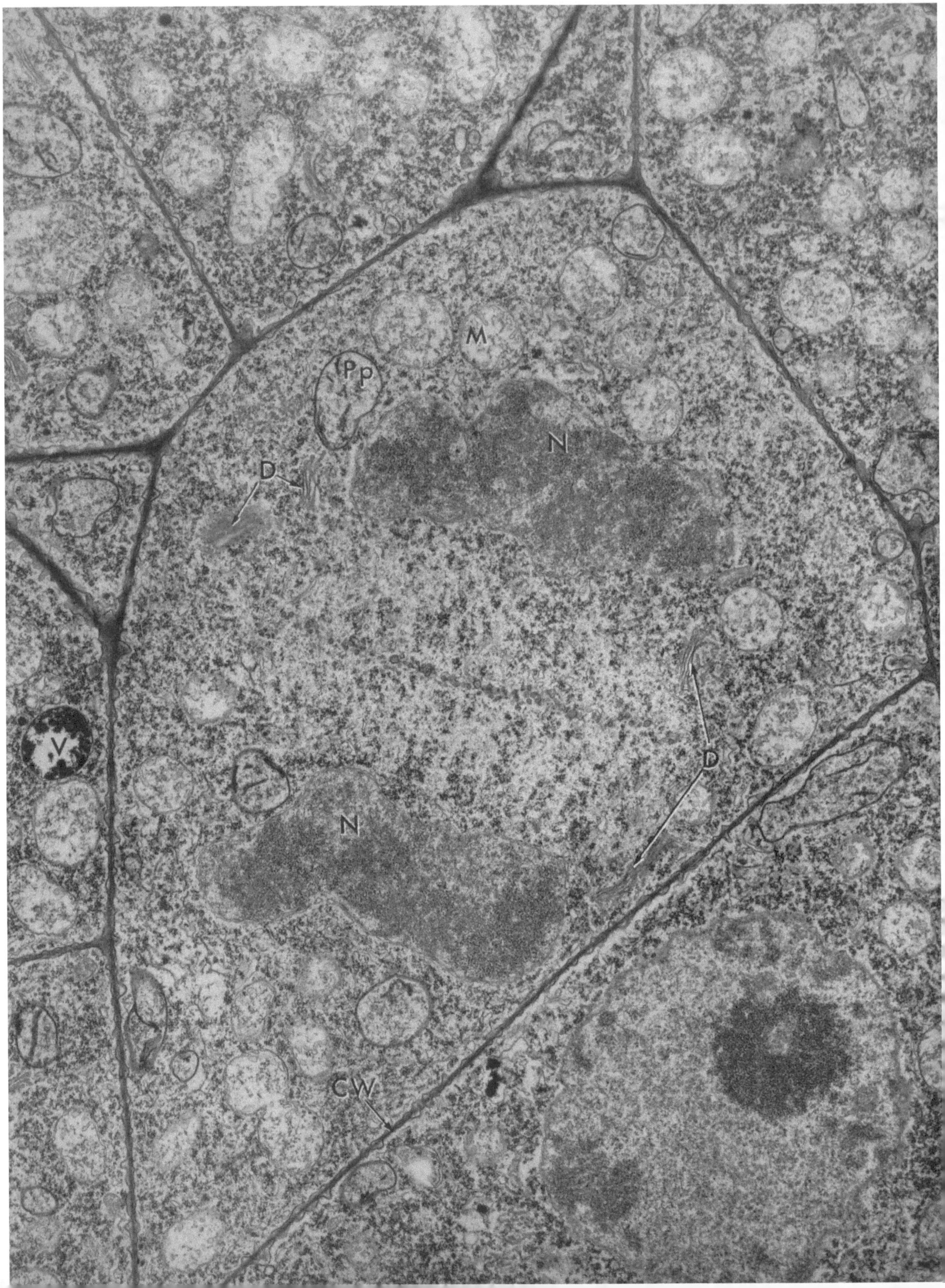

M
Pp
N
D
D
V
N
CW

Plate 3.1

Cytokinesis and Cell Plate Formation

One of the more important phenomena of cells is that of cytokinesis. Quite distinct from mitosis, this sequence of events achieves the formation of two daughter cytoplasts from a single parent cell. It begins in early telophase and is characterized by the appearance first of a line (in section) or plate across the cell, usually at a location midway between the two telophase nuclei. This plate, or cell plate as it is called, starts its development at the center and expands from there towards the margins of the dividing cell. Just marginal to the plate one can recognize a zone of organization, a region of birefringence (see below), which also expands as the plate grows. This zone in its entirety comprises a barrel-shaped structure, which has come to be called the phragmoplast. When it and the plate reach the margins of the parent cell, cytokinesis is complete.

This phenomenon, although better understood now that electron microscopy has resolved some of its morphological details, still presents a number of fundamental questions. What, for example, determines where the plate will form? Not infrequently in plant morphogenesis cell division is unequal or asymmetric, with the result that one daughter cell exceeds the other in size and goes on to a separate fate in its differentiation. The first visible expression of this difference is in the asymmetric positioning of the cell plate. Related to this are questions of origin and organization of the phragmoplast and of what role it plays in cytokinesis. These would be particularly interesting to explore in the anticlinal division of long cambial cells where the nuclei go into interphase long before (a day or two) the phragmoplast and plate reach the ends of the parent cell.

The fine structure associated with cytokinesis is shown in some of its aspects in this plate and the associated text Fig. 3.1a. The cell occupying the center of the micrograph is from the sporogenous tissue of the African violet. A nuclear envelope has assembled around the chromosomes of each daughter nucleus (N), and mitosis has obviously entered telophase. At this time or even slightly before, a large number of tiny vesicles, about 100 nm in diameter, appear in a plane which essentially bisects the interzone between the nuclei. The membranes limiting these vesicles have the dimensions and structure characteristic of the plasma membrane (text Fig. 3.1a). As soon as they collect in this narrow zone, the vesicles begin to fuse and by this act eventually achieve a separation of the two daughter cytoplasts. Since dictyosomes (D) are common around the margins of the plate and have as satellites numerous small vesicles of the same nature as those in the plate, it is thought that the plate vesicles are produced in the Golgi complex. By what device they are brought into the plate is not entirely clear, but the phragmoplast is appropriately positioned and designed to perform this role (see below). After fusion of vesicles in the plate a thin layer of pectin-rich material condenses out between the two new plasma membranes and primary wall formation begins. Mitochondria and proplastids do not enter the interzone or phragmoplast and seem already at this stage to have been assigned to one or the other daughter cells. Profiles of ER vesicles appear to accumulate on opposite sides of the plate after its formation has been initiated, but nothing in their structure or behavior suggests their function in the division of the protoplast.

The cytoplasm at the margin of the plate is normally birefringent. This evidence for the presence of fibrous or filamentous structures arranged parallel to each other and normal to the plate is of feeble intensity at first but gains in prominence as the plate develops. Electron microscopy of such material shows, as in text Fig. 3.1a, that this region or zone of the plate is populated by microtubules. These then make up the birefringent, barrel-shaped phragmoplast which, as mentioned above, marks the limit of the expanding plate. While the phragmoplast is expanding and involving more and more tubule protein, the mitotic spindle and its microtubules are disassembling.

Close study of spindle tubules has shown them to overlap only slightly in the plate and not to extend through it. They therefore individually project into one or the other of the daughter protoplasts but not into both. They seem in fact to take their origins from and to be anchored in condensations of cytoplasmic ground substance which appear as the first indicator of where the plate will form. To what in turn these condensations or dense bodies are related in the organization of the two new cells is undetermined.

It was mentioned earlier that the tubules may at least give direction to the movement of cytoplasmic particles if they are not involved more directly in the generation of the motive force. The assumption that they function here in moving specific vesicles into the plate arises from observations and the fact that they are properly positioned and oriented to do the job. They could, it is thought, draw the vesicles into the spaces between the tubules and propel them to the plate.

In some instances where the microtubules cross the plate and where their initiating centers (dense bodies) persist the continuity of the new wall is interrupted and plasmodesmata are formed. This at any rate is a reasonable explanation for the origin of these cell connections and with our present limited knowledge it seems as valid as any so far proposed.

From sporogenous tissue of the African Violet, *Saintpaulia ionantha* Wendl.
Magnification × 17,000

3.1 Supplemental Reading

Bajer, A.: Cine micrographic analysis of cell plate formation in endosperm. Exp. Cell Res. **37**, 376–398 (1965).

— Notes on ultrastructure and some properties of transport within the living mitotic spindle. J. Cell Biol. **33**, 713–720 (1967).

Becker, W. A.: Recent investigations *in vivo* on the division of plant cells. Botan. Rev. **4**, 446–472 (1938).

Frey-Wyssling, A., Lopez-Saez, J. F., Muhlethaler, K.: Formation and development of the cell plate. J. Ultrastruct. Res. **10**, 422–432 (1964).

Hepler, P. K., Jackson, W. T.: Microtubules and early stages of cell-plate formation in the endosperm of *Haemanthus katherinae* Baker. J. Cell Biol. **38**, 437–446 (1968).

Hepler, P. K., Newcomb, E. H.: Fine structure of cell plate formation in the apical meristem of *Phaseolus* roots. J. Ultrastruct. Res. **19**, 498–513 (1967).

Inoue, S.: Organization and function of the mitotic spindle. In: Primitive motile systems in cell biology, p. 549–598 (Allen, R. D. and Kamiya, N., eds.). New York: Academic Press, Inc. 1964.

Ledbetter, M. C., Porter, K. R.: A "microtubule" in plant cell fine structure. J. Cell Biol. **19**, 239–250 (1963).

Manton, I.: Preliminary observations on spindle fibres at mitosis and meiosis in *Equisetum*. J. roy. micr. Soc. **83**, 471–476 (1964).

Pickett-Heaps, J. D.: Further ultrastructural observations on polysaccharide localization in plant cells. J. Cell Sci. **3**, 55–64 (1968).

Pickett-Heaps, J. D.; Ultrastructure and differentiation in *Chara sp*. Australian J. biol. Sci. **20**, 883–894 (1967).
Porter, K. R., Caulfield, J. B.: The formation of the cell plate during cytokinesis in *Allium cepa* L. In: 4th Intern. Conf. on Electron Microscopy, p. 503–507 (Bargmenn, W., Peters, D., and Wolpers, C., eds.). Berlin-Göttingen-Heidelberg: Springer 1960.
— Machado, R. D.: Studies on the endoplasmic reticulum. IV. Its form and distribution during mitosis in cells of onion root tip. J. biophys. biochem. Cytol. **7**, 167–180 (1960).
Roth, L. E., Daniels, E. W.: Electron microscopic studies of mitosis in amebae. II. The giant ameba *Pelomyxa carolinensis*. J. Cell Biol. **12**, 57–78 (1962).
Sinnott, E. W., Block, R.: Division in vacuolate plant cells. Amer. J. Botany **28**, 225–232 (1941).
Tucker, J. B.: Changes in nuclear structure during binary fission in the ciliate *Nassula*. J. Cell Sci. **2**, 481–498 (1967).
Wilson, H. J.: The fine structure of the kinetochore in meiotic cells of *Tradescantia*. Planta (Berl.) **78** 379–385 (1968).

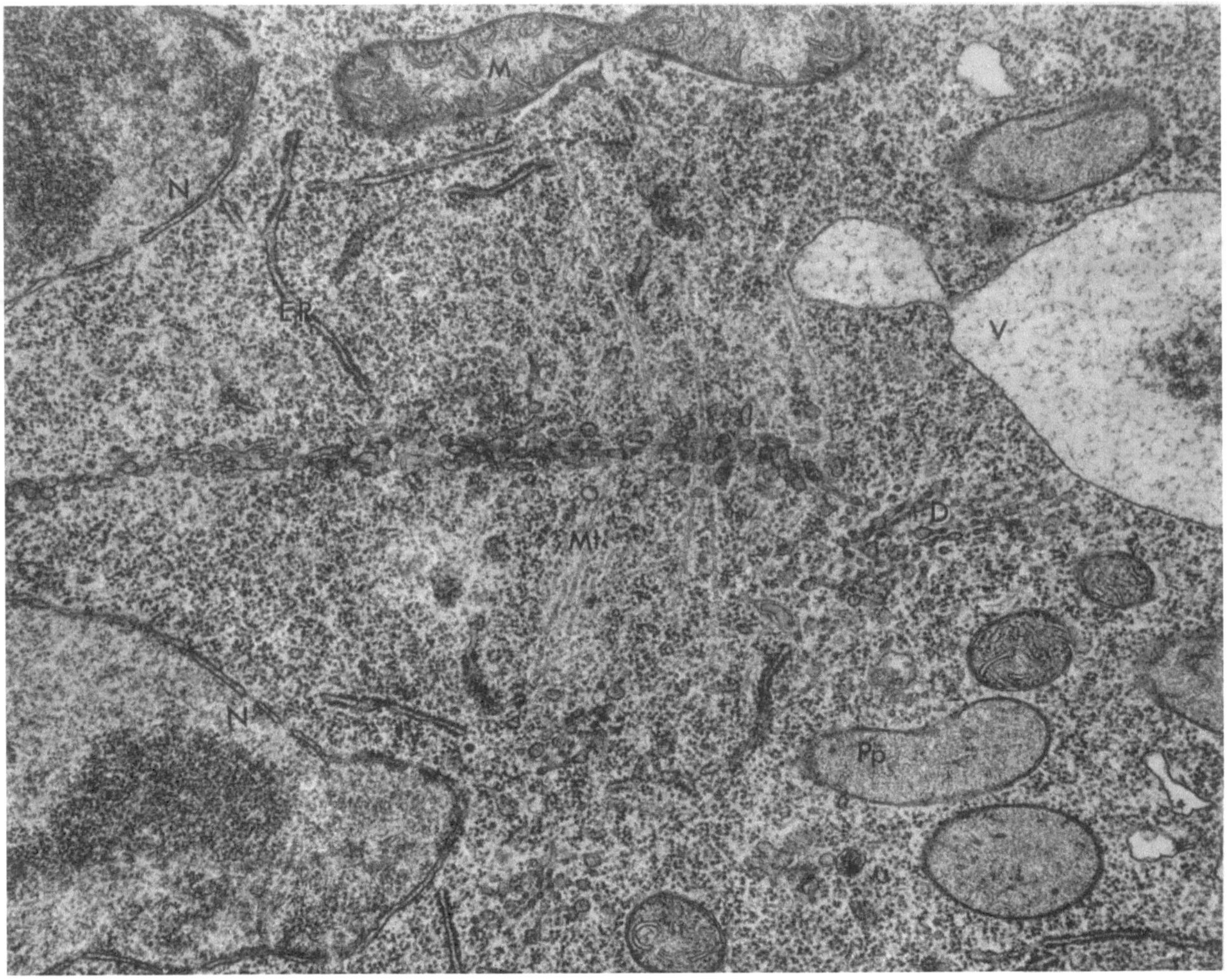
N
M
ER
V
D
Mt
N
Pp

Text Fig. 3.1a

Phragmoplast and Microtubules

This micrograph illustrates microtubules (Mt) as components of the phragmoplast and their relationship to the plate during cytokinesis. The tubules seem in this material to be confined to a marginal zone of the plate. Profiles of the endoplasmic reticulum (ER) enter the interzone and approach the plate from both sides. Ribosomes are also abundant. Proplastids (Pp), mitochondria (M), and the tonoplast-limited vacuole (V) seem, on the other hand, to be excluded. The nuclei (N) are taking on the characteristics of interphase.
From root tip cell of *Arabidopsis thaliana* L.
Magnification $\times$38,000

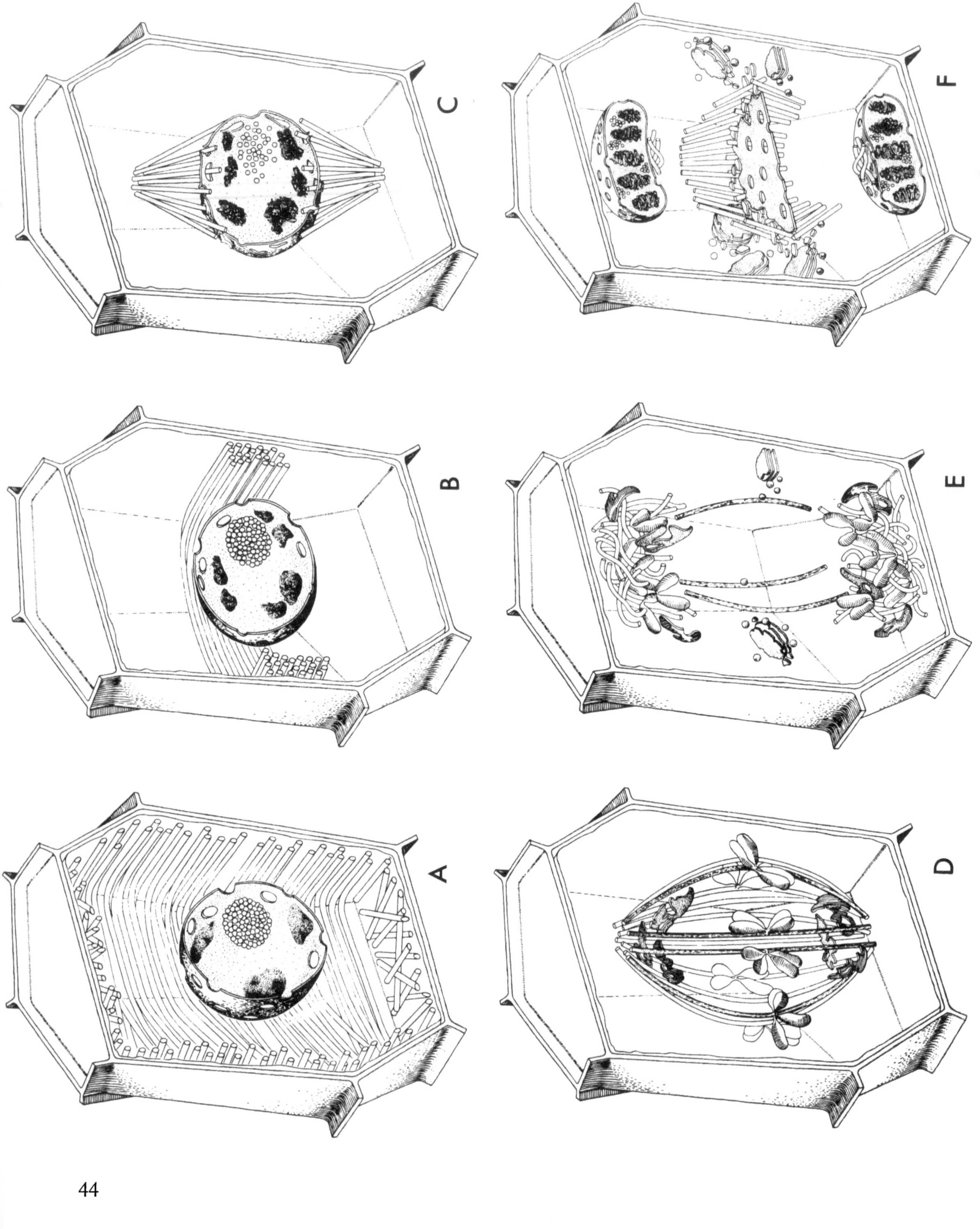
A
B
C
D
E
F

Plate 3.2

Diagram of Changes in Fine Structure during Mitosis

The fibrous structure known as the spindle is a feature of dividing cells of eukaryotic organisms which has attracted much attention since its discovery almost a century ago. Studies of spindle fine structure have shown that the fibers visible in the light microscope are made of bundles of small tubules thought to be identical to the cortical microtubules described in plates 2.4 and 2.4.1. The ubiquitous appearance of microtubules organized in various arrays throughout the mitotic cycle is emphasized in these drawings.

The illustrations represent a cell at various stages of division as it would be seen if cut longitudinally through its central axis to expose the far half to view. The protoplast is encased within the cell wall which it shares with adjacent cells. Only those organelles related directly to the mitotic cycle are illustrated. Some structures are shown somewhat magnified relative to the whole cell to make them easily visible in the diagrams; these include the microtubules, nuclear pores, nucleolar granules, and plasmodesmata. Where information is limited we have made some assumptions which will undoubtedly have to be modified as more evidence accumulates.

The interphase cell (A) is represented with a centrally placed nucleus containing a nucleolus of tightly clustered ribosomal particles. Condensed portions of the chromosomes are contiguous with the nuclear envelope with more diffuse chromatin extending into the nucleoplasm. The nucleus is limited by an envelope of two membranes pierced by numerous pores. Microtubules lie just within the plasma membrane and run circumferentially about the cell axis near the side walls, and at random near the end walls. This organization of microtubules is characteristic of meristematic cells of axial organs such as roots and shoots (Plates 2.4 and 2.4.1). Present information does not permit us to decide whether tubules adjacent to the side walls are in the form of complete loops, spirals of several gyres, or as overlapping arcs. We are equally uncertain of the termination of the randomly arranged tubules near the end walls. It may be that the tubules are in constant state of flux, being assembled from protein subunits at certain loci and disassembled at others.

Condensation of the chromatin and its withdrawal from previous contact with the nuclear envelope are early signs of preparation for division (B). Another mark of the stage which has been termed "preprophase" is the appearance of an equatorial band of 100 to 200 microtubules around the nucleus for some 2.5 μm along the wall. The band is made of several layers of tubules and thus lies deeper within the cytoplasm than the "wall tubules" it replaces. This interesting structure predicts the orientation of the spindle that will develop and the intersection of the succeeding cell plate with the parent wall. When the division is quite asymmetrical as in the development of guard cells of stomata the preprophase band is found along the wall to which the wall of the lenticular guard cell will be attached.

Condensation of the chromatin continues through full-fledged prophase (C). The nucleolus becomes diffuse as its particles disperse into the nucleoplasm. Ruptures appear in the nuclear envelope, probably by enlargement of the nuclear pores. Changes in the microtubules are considerable, as the preprophase band is lost and caps of microtubules appear at the poles on an axis roughly perpendicular to the plane previously occupied by the equatorial band. These cap tubules define a "clear zone" from which larger organelles are excluded. Division in most plant cells is said to be anastral referring to the lack of centrioles and astral rays typical of animal cells. The microtubules grow from these somewhat ill-defined centers at one pole and extend in the direction of the other. Eventually tubules extend also from the kinetochores of the chromosomes in the direction of the poles. The nuclear envelope is disrupted as the spindle of microtubules is established.

The nucleolus is usually undetected at metaphase (D), its particles having dispersed into the cytoplasm. Elements of endoplasmic reticulum are found between microtubules of the polar caps probably as trapped remnants of the disrupted nuclear envelope. The chromosomes, which have now reached their maximum condensation, are aligned with their kinetochores along a plane midway between the poles. One set of tubules, the chromosome microtubules, extends from the kinetochores to the poles and another, the continuous tubules, runs more or less from pole-to-pole. Current thinking is that mechanochemical bridges form between adjacent microtubules in such a way as to orient the chromosomes at the metaphase plate located equidistant from the two poles. The forces of bridging and microtubule growth are thought to bring the spindle to a static equilibrium at metaphase.

Separation of the chromosomes at anaphase (E) has been intensively studied for several decades. It is now thought that this stage is initiated when the equilibrium of metaphase is broken either by weakening of the binding between kinetochores or by increased action of the bridges. The chromosomes are then pulled away from one another by the sliding action of the kinetochore tubules over the continuous tubules of the spindle. According to this scheme, energy is provided by the enzymatic release of high energy phosphate within the bridges to move the tubules in a sliding action analogous to the behavior of filaments of muscle during contraction. Remnants of the kinetochore tubules remain at the poles for a short time. Some of the continuous microtubules are left behind in the interzone between two sets of chromosomes, now separated. The interzone also displays a rich content of ribosomes (not shown) often in conspicuous polysome clusters. Dictyosomes persist throughout the mitotic cycle as do other cytoplasmic organelles, but are shown here because of the role they play in plate formation. Typically the dictyosomes cluster adjacent to the interzone near its equator. Small vesicles which probably contain pectic substances are produced by the dictyosomes and are thought to fuse to form the cell plate.

The nuclear envelope is reconstituted as the clustered chromosomes become wrapped in portions of the endoplasmic reticulum at telophase (F). Pores appear in the cisternae of this system where contact is made with chromosomal material. Dense granules appear along the surface of the chromosomes, not yet fully uncoiled, and these granules later fuse to form one or more nucleoli. A new array of microtubules, constituting the phragmoplast, appear in the interzone between the two daughter nuclei. The plate which cleaves the protoplast in two develops within the phragmoplast. The microtubules appear to overlap slightly at the level of the plate but do not traverse this plane for any considerable distance. The phragmoplast has its counterpart in the stem-body in animal cells. In plant cells it is thought that pectin-containing vesicles produced by the dictyosomes become attached to microtubules and are then guided to their point of fusion at the margin of the plate. Plasmodesmata (Section 4) appear in some unknown manner in the developing plate. The phragmoplast expands as the plate becomes larger until the wall of the parent cell is reached. The resulting two daughter cells will enter interphase (A) to complete mitosis and cytokinesis.

3.2 Supplemental Reading

Burgess, J., Northcote, D. H.: A function of the preprophase band of microtubules in Phleum pratense. Planta (Berl.) **75**, 319–326 (1967).

Cronshaw, J., Esau, K.: Cell division in leaves of Nicotiana. Protoplasma **65**, 1–24 (1968).

Hepler, P. K., Jackson, W. T.: Microtubules and early stages of cell-plate formation in the endosperm of *Haemanthus katherinae* Baker. J. Cell Biol. **38**, 437–446 (1968).

Ledbetter, M. C.: The disposition of microtubules in plant cells during interphase and mitosis. Symp. Intern. Soc Cell Biol. **6**, 55–70 (1968).

McIntosh, J. R., Hepler, P. K., Wie, D. G. van: Model for mitosis. Nature (Lond.) **224**, 659–663 (1969).

Pickett-Heaps, J. D., Northcote, D. H.: Cell division in the formation of the stomatal complex of the young leaves of wheat. J. Cell Sci. **1**, 121–128 (1966).

— — Organization of microtubules and endoplasmic reticulum during mitosis and cytokynesis in wheat meristems. J. Cell Sci. **1**, 109–120 (1966).

Wilson, G. B.: Cell division and the mitotic cycle. 111 pp. New York: Reinhold Publ. Corp. 1966.

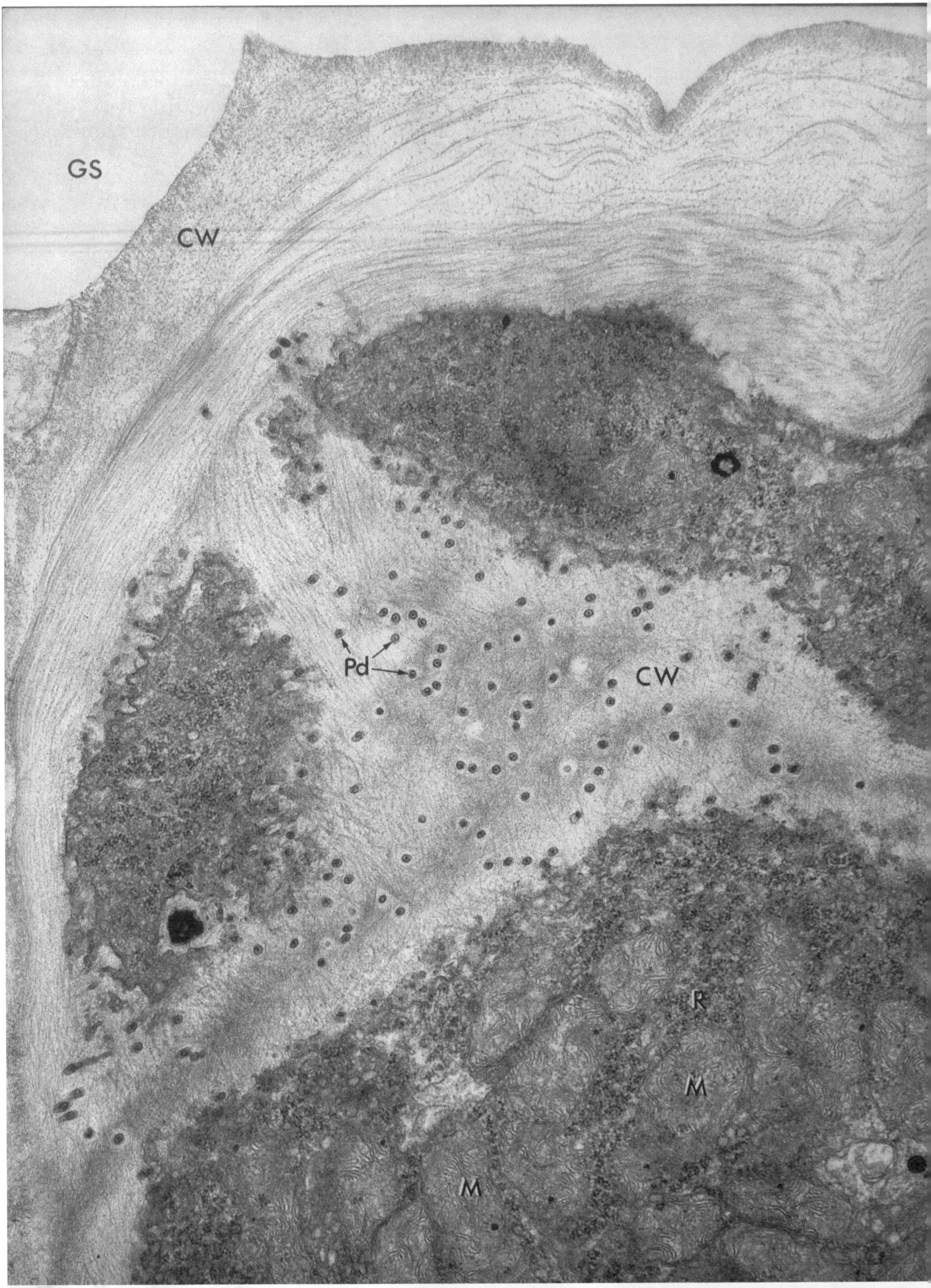
GS
CW
Pd
CW
R
M
M

Plate 4.1

Plasmodesmata between Cells

Virtually all living cells of the higher plant are interconnected by numerous slender protoplasmic strands or bridges called plasmodesmata. These make of the plant a continuous protoplast, a kind of syncytium invaded extensively by cellulose walls, which lend physical support to the highly varied and extraordinary forms that plants adopt. Though they are individually small, the plasmodesmata provide for the intercellular movements of large and small molecules, and for the transport from cell to cell of particles as large as viruses. Presumably they offer a barrier to some forms of genetic information, for otherwise differentiation among the cells of the population making up the plant would be impossible. It is doubtless of some significance that the plasma membranes are continuous through the plasmodesmata, possibly providing for a widespread equilibration of membrane potentials and the transfer of membrane-supported excitations. With such varied and important functions assigned to them these structures are a unique and highly interesting feature of plant morphology.

Plasmodesmata (Pd) are shown here as they exist in the end walls of cells making up the wheat stamen filament. The section, cut transversely with respect to the filament, includes a portion of the wall (CW) and parts of two adjacent cells, one at the bottom and the other, in two parts, at the top and left. A thick wall characteristic of these collenchymal cells is shown in the upper left quadrant adjacent to an intercellular space (GS).

It is evident that the plasmodesmata are remarkably uniform in diameter and that each consists of a dense circumference with a small dense central element. Each is then tubular in form and about 400 Å in diameter. As shown more clearly in plates 4.1.1 and 4.1.2, the limiting wall of the tubule is a unit membrane and is continuous with that of the adjacent cells. Although they are nearly always present, the number of these intercellular connections varies enormously from time to time. They are notably absent, however, from walls separating generative cells (sporogenous tissue) from each other and from the surrounding tapetal cells of the sporangium (see Plate 10.2).

If one were to take seriously the speculation on plasmodesmata one would have to conclude that their origin is varied and probably involves the intervention of more cytoplasmic components than is generally realized. It has been suggested that some plasmodesmata represent strands of cytoplasm harbouring vesicles of the ER and microtubules that are included in the cell plate during cytokinesis. Others, however, seem to be established between cells in interphase through some mechanism of localized wall breakdown and reorganization. It is well known, for example, that they develop between cells which are brought anew into contact, as between cells in graft unions and between haustoria of parasite and cells of host.

The extraordinary capacity of wheat filament cells to grow at a rapid rate lends particular interest to this micrograph. Several investigators have reported that through elongation of the cells this filament may extend 2–3.5 mm per minute. To allow for this there is at an earlier stage of morphogenesis a rapid production and assembly of wall materials. Both processes require simultaneous production of enzymes and ATP, and all activities are doubtless coordinated in all cells of the filament by the abundant supply of plasmodesmata.

The wall of these cells is seen to be a loose arrangement of fine filaments separated by large spaces probably containing a substantial proportion of water. Thus the walls retain a plasticity essential for the rapid changes in shape during filament elongation. The cellulose filaments are obviously parallel and circumferentially arranged in the outside wall but randomly oriented between the plasmodesmata in the cross walls.

The protoplast surface, at the plasma membrane, is very irregular. This is characteristic of other known plant cell surfaces adjacent to sites of rapid wall deposition. The large populations of mitochondria (M) with numerous cristae are properly interpreted as an adaptation for the rapid production of energy-rich compounds close to the sites where they are required. The dense concentrations of ribosomes (R), many in polysomal aggregations, are properly associated with the production of more of the intracellular structural proteins and/or the synthesis of polymerases involved in wall formation. For the short duration of its adult life, this cell is one of the most active to be found in any plant tissue.

From the anther filament of wheat, *Triticum aestivum* L.

Magnification × 34,000

4.1 Supplemental Reading

Buvat, R.: L'infrastructure des plasmodesmes et la continuite des cytoplasmes. C. R. Soc. Biol. (Paris) **245**, 198–201 (1957).

Green, P. B.: On mechanisms of elongation. In: Cytodifferentiation and macromolecular synthesis, p. 203–234 (Locke, M., ed.). New York: Academic Press, Inc. 1963.

Juniper, B. E.: Origin of plasmodesmata between sister cells of the root tips of barley and maize. J. roy. micr. Soc. **82**, 123–126 (1963).

Meeuse, A. D. J.: Plasmodesmata. Botan. Rev. **7**, 249–262 (1941).

Miller, E. C.: Plant physiology. 2nd ed. New York: McGraw-Hill 1938.

Ohad, I., Danon, D.: On the dimensions of cellulose microfibrils. J. Cell Biol. **22**, 302–305 (1964).

Robards, A. W.: A new interpretation of plasmodesmatal ultrastructure. Planta (Berl.) **82**, 200–210 (1968).

Roelofsen, P. A.: The plant cell-wall. Berlin-Nikolassee: Gebrüder Borntraeger 1959.

Scott, F. M., Bystrom, B. G., Bowler, E.: *Persea americana*, mesocarp cell structure, light and electron microscope study. Botan. Gaz. **124**, 423–428 (1963).

Wardrop, A. B.: Cell wall organization in higher plants. I. The primary wall. Botan. Rev. **28**, 241–285 (1962).

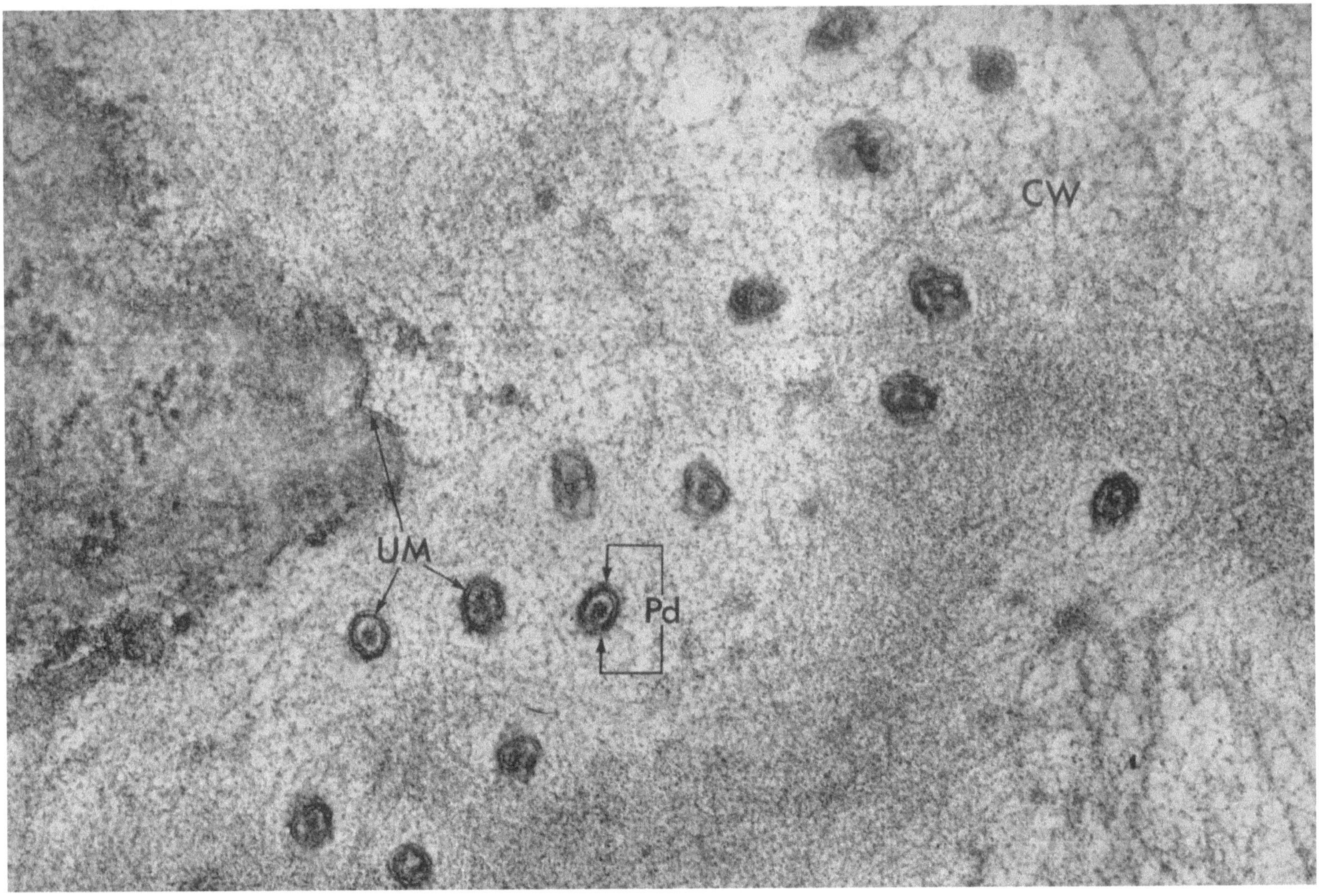
CW
UM
Pd

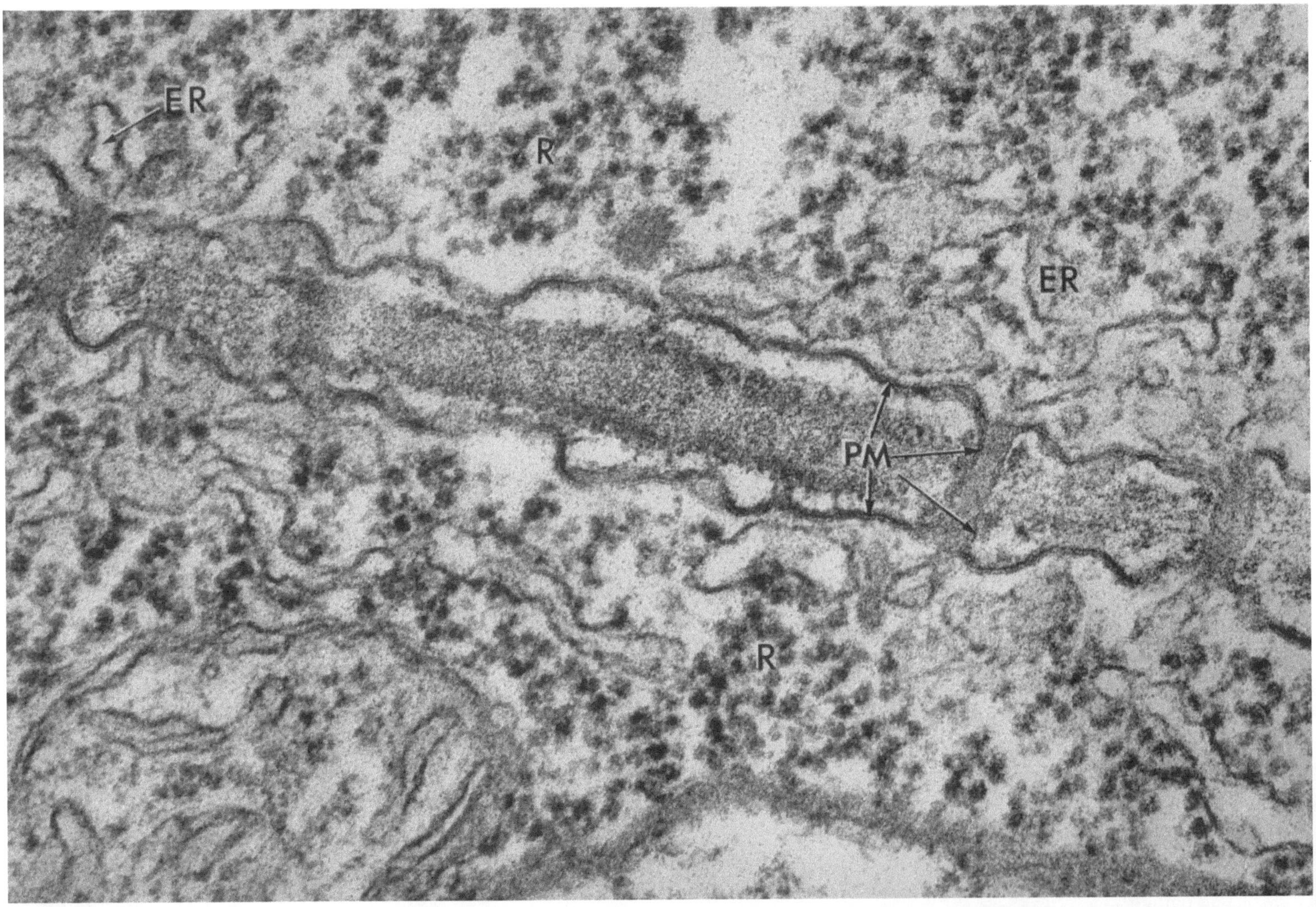
ER
R
ER
PM
R

Plates 4.1.1 and 4.2.2

Details of Plasmodesmata

The ubiquity of plasmodesmata in the living tissues of higher plants indicates the functional importance of these structures. Even so, their full significance is probably still not appreciated. In recent years morphological studies have made a number of valuable contributions to the development of new or to the support of old hypotheses that are now being reinvestigated.

These two micrographs collectively depict the major features of plasmodesmata substructure. The cross sections shown in Plate 4.1.1 (upper) are photographic enlargements of the plasmodesmata in an area of Plate 4.1. The increased magnification makes more easily visible the trilaminar structure of the unit membrane (UM) limiting the plasmodesmata. This has exactly the same dimensions and structure as the plasma membrane. The cross sections of the plasmodesmata also show, within each lumen, a clear zone about 100 Å in width, surrounding a central dense core of unknown nature.

The longitudinal views in 4.1.2 (lower) expand slightly on this interpretation. Since the plasmodesmata are only 400 Å in overall diameter and since most "thin" sections are 500 Å or more in thickness, it becomes difficult to get a completely informative longitudinal view. Nevertheless this section includes at least two half-plasmodesmata, essentially as if they were cylinders split down their long axes. Clearly the limiting membrane is continuous with those of both protoplasts. But neither of these two plates illuminates the core structure. If derived in the first instance from the ER, it has come to be compressed by the narrowness of the pore into something no longer recognizable as ER membrane or tubule. What seems equally plausible is that the material of the plug is a condensation of the cytoplasmic ground substance. It is true that profiles of ER vesicles usually approximate closely the ends of the plasmodesmata, but structural continuity through the plasmodesmata is not demonstrable. Where, as here, ribosomes (R) and ER profiles (ER) are clearly depicted, one can see how relatively difficult it would be for a polysome plus messenger RNA to move through one of these channels. Anything membrane-borne, however, and moving over the continuous cell surface, would seemingly have no choice but to follow the limits of the plasmodesmata.

It is characteristic of the plasmodesmata profile in longitudinal section to show a slight outpocketing at the level of the middle lamella. This may reflect a softening of the middle lamella relative to the layers of primary wall (cellulose) on either side, and consequently a bulging out of the membrane at that level.

4.1.1 from wheat anther filament, *Triticum aestivum* L.

4.1.2 from wheat leaf, *Triticum aestivum* L.

Magnifications 4.1.1 × 130,000
Magnifications 4.1.2 × 60,000

4.1.1 and 4.1.2 Supplemental Reading

Dalyman, P.: Elektronenmikroskopische Untersuchungen an den Saughaaren von *Tillandsia usneoides* (Bromeliaceae). II. Einige Beobachtungen zur Feinstruktur der Plasmodesmen. Planta (Berl.) **64**, 76–80 (1965).

Dolzmann, R., Dolzmann, P.: Untersuchungen über die Feinstruktur und die Funktion der Plasmodesmen von *Volvox aureus*. Planta (Berl.) **61**, 332–345 (1964).

Franke, W.: Ektodesmenstudien. III. Mitteilung zur Frage der Struktur der Ektodesmen. Planta (Berl.) **63**, 279–300 (1964).

Lopez-Saez, J. F., Gimenez-Martin, G., Risueno, M. C.: Fine structure of the Plasmodesm. Protoplasma (Wien) **61**, 81–84 (1966).

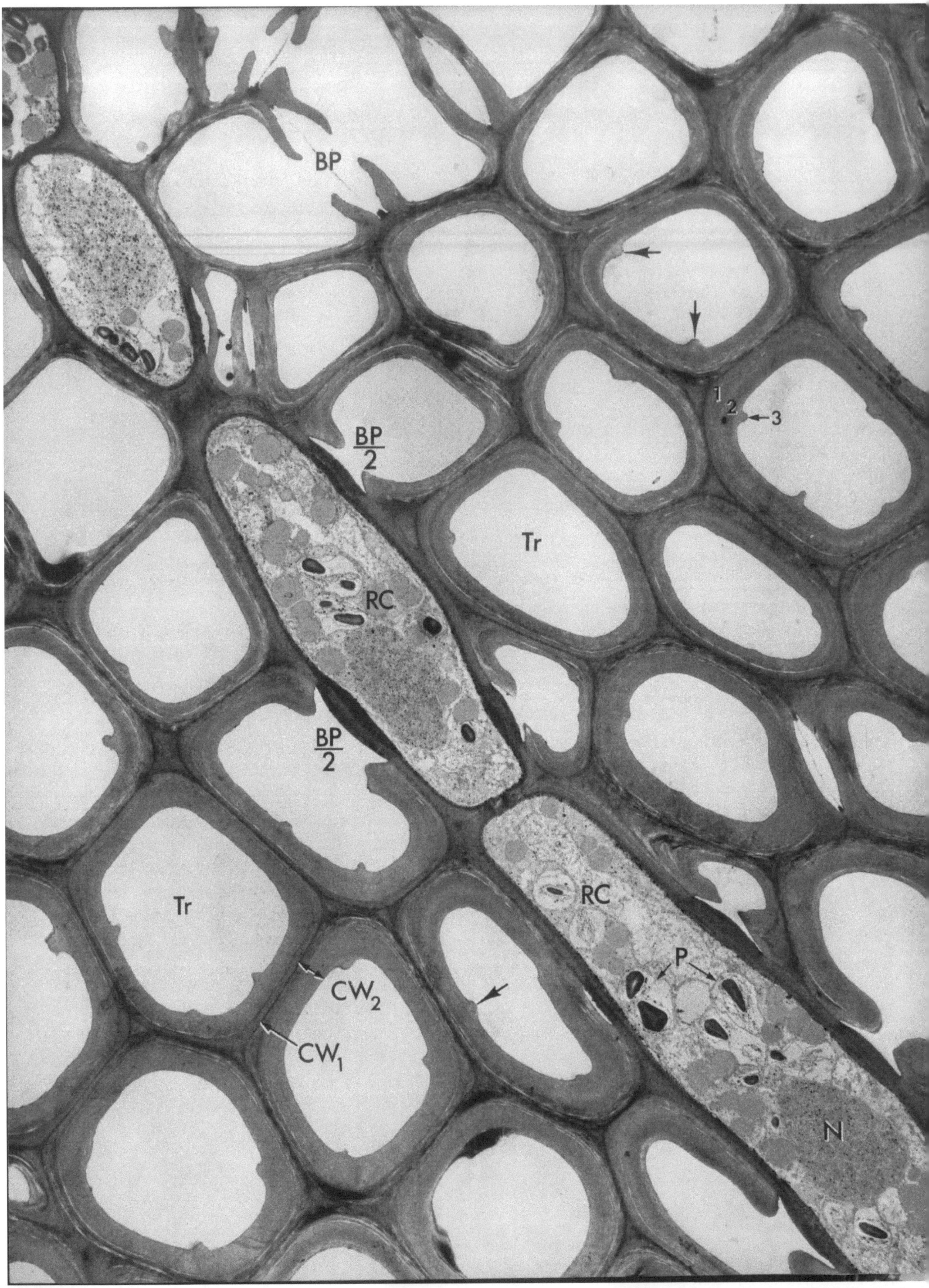

BP
BP/2
1
2
3
Tr
RC
BP/2
RC
Tr
P
CW2
CW1
N

Plate 5.1

Secondary Xylem

The bulk of the woody stems of plants is made up of slender, cylindrical, thick-walled cells called tracheids. These tracheids, together with the larger vessels when present, constitute the channels by which water is moved from the roots to the leaves.

The tracheids (Tr) of *Taxus* are unusually small as such cells go, measuring not more than 10–15 μm in diameter. For this reason, they are especially valuable for electron microscopy. The cross sectional form is obviously cylindrical. Their length exceeds their diameter 80–130 times. It is obvious that tracheids are without living protoplasts. Indeed, it is known that only a small fraction of the original protoplast remains when wall formation is completed, and this residue soon autolyzes. Remnants of the protoplast may contribute finally to waterproofing the wall.

The cellulose walls of these xylem elements as they occur in mature wood vary only slightly in thickness. Here they measure approximately 1.0 μm. The inner surface faces on the lumen of the element, the outer on the middle lamella, sometimes evident as a very thin line between individual cells. In this particular tissue of the yew and at this magnification the middle lamella and the primary walls of adjacent tracheids are indistinguishable. In fact the primary walls (CW_1) blend into a single thin layer between the surfaces of adjacent tracheids and expand into a thicker layer only at the corners to fill what would otherwise be intercellular spaces. This thin primary wall usually appears denser than the rest of the wall in preparations of this kind, possibly because of a greater content of lignin.

The cellulose-rich portion of the wall (CW_2) of the tracheid is put down secondarily to the first or primary layer. It, too, is obviously divided into layers, an outer thin one next to the primary (1) and an inner thicker one (2) (a major part of the wall, called S_2) (see Plate 5.2). These layers are thought to represent phases in type and rate of cellulose synthesis within the original protoplast. Evidences of a third or inner layer of the wall (3) are recognized in the small warts that project into the lumen (arrows).

These extraordinary walls perform several functions. They doubtless give the woody stem its stress-resisting properties and its capacity to support the huge weights represented by the upper branches and parts of the plant. But just as significant is their role in resisting the tensions arising within the long column of water in the stem. The continuity of this column must be maintained under stresses amounting to hundreds of atmospheres of pressure. Clearly the walls of individual, fully differentiated tracheids must be rigid and strong enough to maintain the tracheids as open cylinders against such large negative tensions.

At various points along their side walls adjacent tracheids may establish connections via pores called pits. These form as the tracheids develop and really represent circular zones where the secondary wall fails to be deposited in the usual relationship to the primary one. Sometimes, as here in the locust, the free edge of the secondary wall extends out over the pit area to form a border. In this and similar instances the pits are said to be bordered (BP). In many plants, particularly among the Angiosperms, the marginal cutoff in deposition of secondary wall around the pit is more abrupt, and the pits are described as simple. Ordinarily the primary wall or some modification of it extends across the pit (see Plate 5.2). This diaphragm constitutes the pit membrane, which possesses varying degrees of porosity in different plants. The movement of water and nutrients through this system of tracheids is primarily via the pits. The vessel elements, as found in most Angiosperms and several other groups, may have pits along their side walls, but are more directly connected to one-another by end wall perforations. These perforations are relatively broad openings free of any obstructing remnants of the primary wall; thus, vessel elements impose less resistance upon the flow of xylem sap than do tracheids. Where a tracheid is adjacent to a ray cell (RC), it develops pits that are bordered only on the tracheid side and are therefore referred to as half bordered (BP/2). Here, as in the intertracheid pits, the pit membrane on the tracheid side is continuous with the primary wall only.

The ray cells (RC), just mentioned, are shown in this plate as a single file passing across the picture from upper left to lower right. They are, in comparison with tracheids, fairly small cells, being 20–30 μm in their longest dimension. Identified and studied much earlier by light microscopy, the ray cells are known to be arranged in sheets or bundles, which radiate out from the center of the stem. Unlike other cells in the xylem these are living and remain so throughout the functioning life of the sap wood. This is evidenced in this micrograph by the presence of a nucleus (N), plastids with starch grains (P), and mitochondria. Other inclusions, which are spherical in outline and have uniformly dense material within are thought to be protein-storage granules. The starch granules acquire a pronounced density following the preparation techniques employed here. Each ray cell is surrounded by a thick primary wall, which is continuous with the primary walls of the adjacent tracheids. No secondary wall of the usual type is evident, although next to the protoplast there is a layer of dense material that thickens somewhat over the half-bordered pit. At their ends, where they abut on other ray cells in the file, the primary wall is traversed by plasmodesmata.

The function of ray cells is only partly understood. They are generally regarded as providing for transport transversely in the stem and particularly from xylem to phloem. It is evident as well that they engage in storage of carbohydrate and protein. One famous example of this activity comes from studies on sugar maples showing that the sugar content of the spring sap is derived from the starch stores of the ray cells.

From a stem wood of *Taxus canadensis* Marsh. Magnification $\times$ 3,800

5.1 Supplemental Reading

Albersheim, P.: The substructure and function of the cell wall. In: Plant biochemistry p. 151–185 (Bonner, J., and Varner, J. E., eds.). New York: Academic Press 1965.

Bailey, I. W.: Evolution of the tracheary tissue of land plants. Amer. J. Botany **40**, 4–8 (1953).

Cronshaw, J.: Cytoplasmic fine structure and cell wall development in differentiating xylem elements. In: Cellular ultrastructure of woody plants p. 99–124 (Cote W. A., Jr. ed.). Syracuse, N.Y.: Syracuse University Press 1965.

— The formation of the wart structure in tracheids of *Pinus radiata*. Protoplasma (Wien) **60**, 233–242 (1965).

— Tracheid differentiation in tobacco pith cultures. Planta (Berl.) **72**, 78–90 (1967).

Liese, W.: The warty layer. In: Cellular ultrastructure of woody plants p. 251–269 (Cote, W. A., Jr., ed.). Syracuse, N.Y.: Syracuse University Press 1965.

Mühlethaler, K.: Ultrastructure and formation of plant cell walls. Ann. Rev. Plant Physiol. **18**, 1–24 (1967).

Pickett-Heaps, J. D., Northcote, D. H.: Relationship of cellular organelles to the formation and development of the plant cell walls. J. exp. Botany **17**, 20–26 (1966).

Porter, K. R., Machado, R. D.: The endoplasmic reticulum and the formation of plant cell walls. In: Proc. European Regional Conf. Electron Microscopy, Delft, vol. 2, p. 754–758 (Houwink, A. L., and Spit, B. J., eds.). Delft: De Nederlandse Vereniging voor Electronenmicroscopie 1961.

Ray, P.: Radioautographic study of cell wall deposition in growing plant cells. J. Cell Biol. **35**, 659–674 (1967).

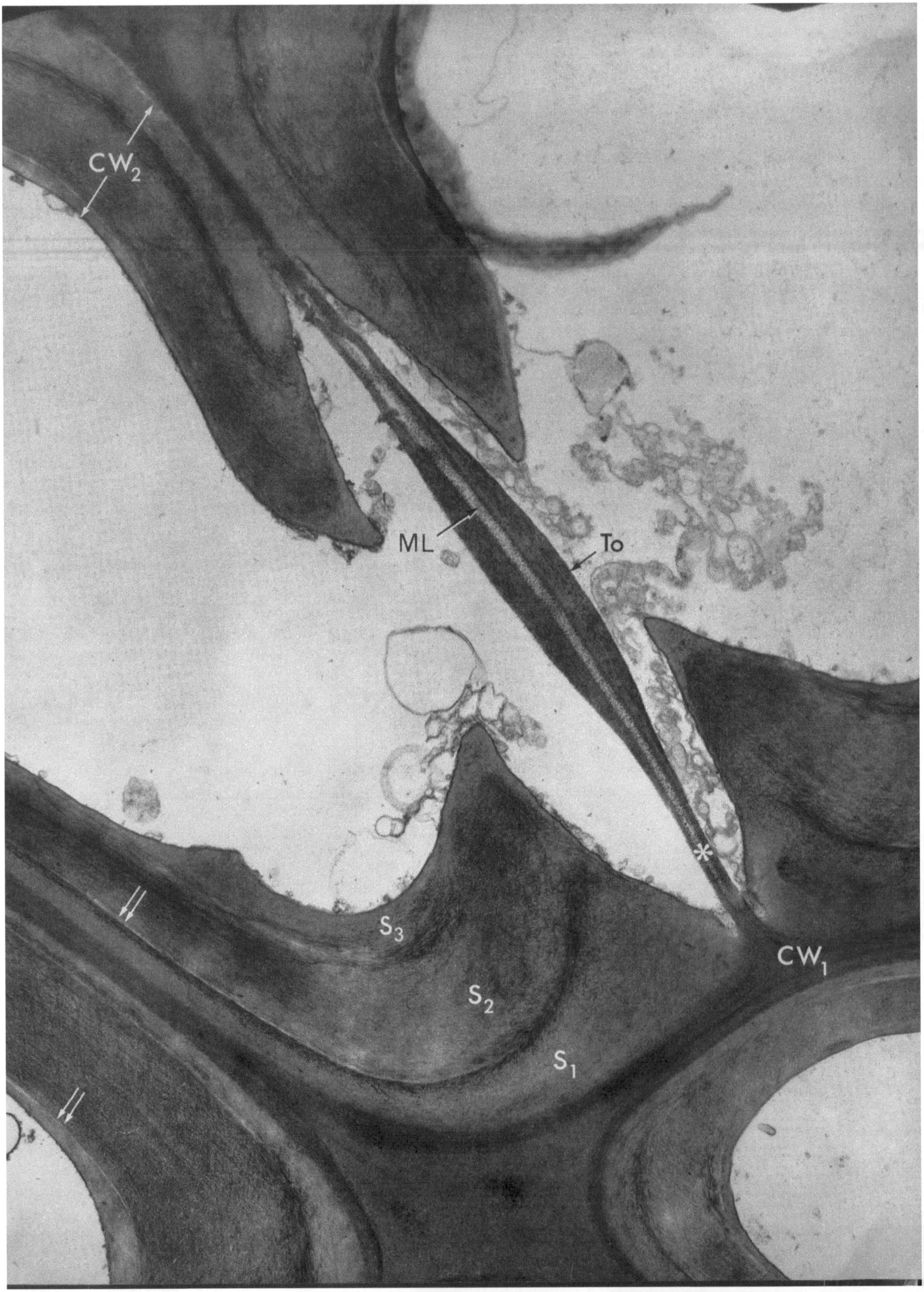
CW2
ML
To
*
S3
S2
S1
CW1

Plate 5.2

Bordered Pit

As mentioned in relation to Plate 5.1, pit-pairs represent special channels for the passage of water and nutrients between the thick-walled cells of woody stems. They are commonly found between tracheids or between tracheids and vessels of the xylem.

The immature bordered pit-pair shown here in profile was found between two tracheids which, when fixed, were in the final stages of differentiation. Only remnants of the intact protoplasts, earlier present, remain. They appear as isolated vesicles with surrounding amorphous materials in stages of disintegration.

This pit-pair comprises three major parts: (a) the pit aperture, which is simply an interruption in the continuity of the secondary wall (CW_2), (b) the pit membrane (*), which extends like a diaphragm across the pit, and (c) the overhanging margins of the secondary wall which gives each pit its border.

The pit membrane is obviously the most significant part of the structure, since its properties determine what passes between the cells and at what rate. A central layer within the membrane is less dense than the rest of the structure and represents a remnant of the original middle lamella (ML). It extends only as far as the margins of the pit. The thicker and denser layers of the membrane on opposite sides of the ML are derivations of the original primary walls of the tracheids. Clearly these layers are continuous at the margins of the pit with the dense primary wall (CW_1) surrounding the tracheids. At its center the pit membrane thickens into a zone that is called the torus (To) and is more than large enough in diameter to cover the pit aperture should the membrane be forced to one side or the other within the pit.

The overhanging lips bordering the pit are, as mentioned before, extensions of the secondary wall (CW_2). Here and elsewhere it is evident that this layer in the structure of the tracheid wall is not homogeneous. Instead it is layered, with a characteristic fine structure within each layer. As an aid to discussion, these layers have been designated as S_1, S_2 and S_3.

The S_1 layer, next to the primary wall, has itself a microfibrillar, layered structure. It is known from several studies that the microfibrils run helically around the cell at a shallow pitch relative to the transverse plane. In the S_2 layer, which is the thickest, the microfibrils also run helically but shift their orientation to a relatively steep angle of 60–70° relative to the transverse. It is said to contribute the greatest tensile strength to the wall. The innermost layer, the S_3, is the thinnest of all and shows little or no evidence of a microfibrillar substructure. Where its substructure has been exposed, the microfibrils of cellulose are shown to be oriented as in the S_1 layer. Apparently the S_3 layer has a higher content of matrix substances (pectins, lignins, sporopollenins) which give it a much greater resistance to chemical attack. That these various layers or plies are distinct entities, laid down separately by a protoplast that had changed its organization and metabolic activity, is indicated by the evident tendency of the layers to separate (double arrows). The bonding is obviously weaker between than within layers.

There is evidence that the thinner marginal portion of the pit membrane is constructed to permit relatively free passage of water and solutes from tracheid to tracheid. In the pits of hardwood tracheids and vessels the membranes are uniform throughout and without prominent substructure. In those of conifers on the other hand, the central part of the membrane, i.e. the torus, when fully differentiated, is suspended by a marginal portion which appears as a sieve or network of fibrils (see Plate 5.2.1). Compared with the torus, then, which is dense and homogeneous in structure, the peripheral zones of the pit membranes are highly permeable. This variation within the structure of the conifer pit membrane appears to have a special purpose especially in older stems, in which occasionally, the water column in some tracheids will break with the development of a gas bubble. When the bubble occurs opposite a bordered pit, it forces the torus over against the border. Thus the thickened torus functions like a valve to close the pit and prohibit the passage of gas to the next tracheid, where it would break the water column and further impede normal transport. Once moved from its central free position to the lips of the border, the torus may become cemented in place, permanently closing the pit.

From[1] the woody stem of *Taxus canadensis* Marsh.

Magnification × 19,000

[1] Reprinted from Botany: An Introduction to Plant Biology, 4th ed., by Weier, T. E. Stocking, R. and Barbour, M. G., New York: John Wiley and Sons, Inc. 1970.

5.2 Supplemental Reading

Cronshaw, J.: The fine structure of the pits of *Eucalyptus regnans* (F. Muell.) and their relation to the movement of liquids into the wood. Austr. J. Botany **8**, 51–57 (1960).

Liese, W.: Zur Struktur der Tertiarwand bei den Laubhölzern. Naturwissenschaften **44**, 240–241 (1957).

— Bauch, J.: On the closure of bordered pits in conifers. Wood Sci. Technol. **1**, 1–13 (1967).

— — Scholz, F.: Über den Torus der Hoftüpfel in Coniferenholz. Naturwissenschaften **54**, 49 (1967).

Machado, R. D., Schmid, R.: Observations on the structure of pit membranes in hardwood cells. In: Proc. Third European Regional Conf. Electron Microscopy, Prague, p. 163–164 (Titlbach, M., ed.). Prague: Publishing House of the Czechoslovak Academy of Sciences 1964.

Schmid, R.: The fine structure of pits in hardwoods. In: Cellular ultrastructure of woody plants, p. 291–304 (Côté, W. A., Jr. ed.). Syracuse, N.Y.: Syracuse University Press 1965.

— Machado, R. D.: Über den Feinbau der „verzierten" Tüpfel bei der Gattung *Plathymenia*. Holz, Roh-Werkstoff **21**, 41–47 (1963).

— — Zur Entstehung und Feinstruktur Skulpturierter Hoftüpfel bei Leguminosen. Planta (Berl.) **60**, 612–626 (1964).

Schmid, Rudolph: Electron microscopy of wood of *Callixylon* and *Cordaites*. Amer. J. Botany **54**, 720–729 (1967).

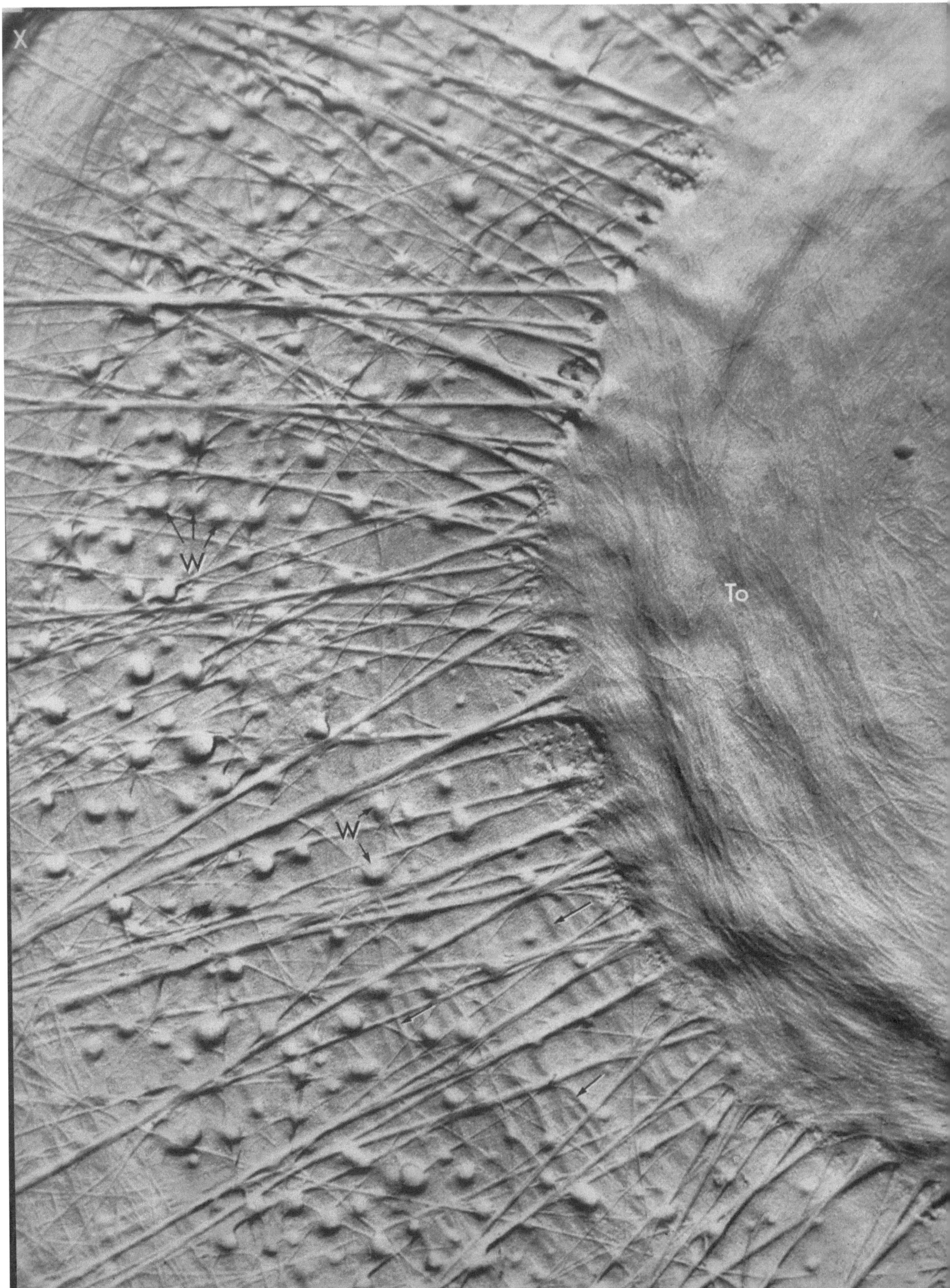
X
W
W
To

Plate 5.2.1

Pit Membrane and Torus

Thin sections of plant or animal tissues are essential for an understanding of many aspects of fine structure, but they fail in at least one respect and that is to provide a good clear image of the cell or tissue surface. Fortunately other techniques have been developed for this latter purpose, techniques that succeed quite simply in producing a replica of those surfaces. Such an approach has been used to obtain face views at high resolutions of cellulose walls and of pit membranes, as shown here.

A short description of the procedure will help the viewer interpret this image. When a woody stem is split, the natural cleavage lines that the split follows usually coincide with the middle lamellae between the cells. Where pits are involved, the pit cavity is opened up but the pit membrane remains with one or the other half. When this exposed surface is firmly pressed against a thermally softened plastic, a negative replica of the wood surface is obtained. This can be "shadowed" by subliming metal such as platinum over it at a low angle (under vacuum) in order to define and accentuate the elevations and depressions in the surface. Finally the metal positive replica so produced is backed up with an evaporated layer of amorphous carbon. When removed from the original plastic replica, it is ready for study under the electron microscope. The elevations in the original material give the illusion in the electron micrograph of being illuminated by a low light from some source above the field of the image.

About one half of a pit membrane as found in the bordered pits of this conifer is shown in this plate. The centrally located torus is included at the right (To); the pit margin (X) appears in the picture at the upper left. The replica in between reproduces the fine structure of the intervening pit membrane.

Clearly the torus (To) in this pit membrane is a disc-shaped pad occupying the center of the membrane. It is constructed of innumerable microfibrils of cellulose each about 100 Å in diameter and arranged in obvious relation to the contours of the torus, especially at its margin. Beyond this margin the pit membrane changes its character quite dramatically. It becomes a meshwork of thick strands and unit fibrils of cellulose. The thicker elements (bundles of microfibrils) extend radially from the torus to the limits of the pit. These strands, 100 nm thick and smaller, are made up of several microfibrils and seem to originate deep within the torus. After branching at various levels, they continue into the primary wall around the pit. Much finer microfibrils intermingle randomly with these radial fibers, except near the outer edges of the pit membrane where they are oriented mostly circumferentially, or parallel to this edge. Thus the pit membrane beyond the torus is a meshwork of fibrils with fairly large openings (0.1–0.5 μm) between the strands. The porosity of this membrane has been examined and shown to permit the passage of colloidal gold particles with diameters as large as 100 to 200 nm, a size that matches the pores in the membrane. Obviously such a "membrane" would offer far less resistance to the movement of xylem sap from cell to cell than would the more tightly constructed immature pit membrane present in *Taxus* (Plate 5.2).

In this image the surface which lies beyond or behind the pit membrane is the inner surface of the border, created by the overhang of secondary and tertiary cell wall. It is warty as evidenced by the small rounded elevations (W), which protrude into the plane of the membrane. Except for the warts, it is relatively amorphous as would be expected for the surface of the tertiary or S_3 layer of the wall that covers this surface. A circumferential ridging of this surface (arrows) suggests that the organization of the tertiary layer does bear some relation to the circular outline of the pit. From[1] the woody stem of *Pinus sylvestris* L. Magnification × 20,000

[1] This micrograph was generously provided by Walther Liese.

5.2.1 Supplemental Reading

Bailey, I. W.: The structure of the bordered pits of conifers and its bearing upon the tension hypothesis in the ascent of sap in plants. Botan. Gaz. **62**, 133–142 (1916).

Harris, J. M.: Heartwood formation in *Pinus radiata* (D. Don.) New Phytologist **53**, 517–524 (1954).

Liese, W.: The fine structure of bordered pits in softwoods. In: Cellular ultrastructure of woody plants, p. 271–290 (Côté, W. A., Jr., ed.). Syracuse, N.Y.: Syracuse University Press 1965.

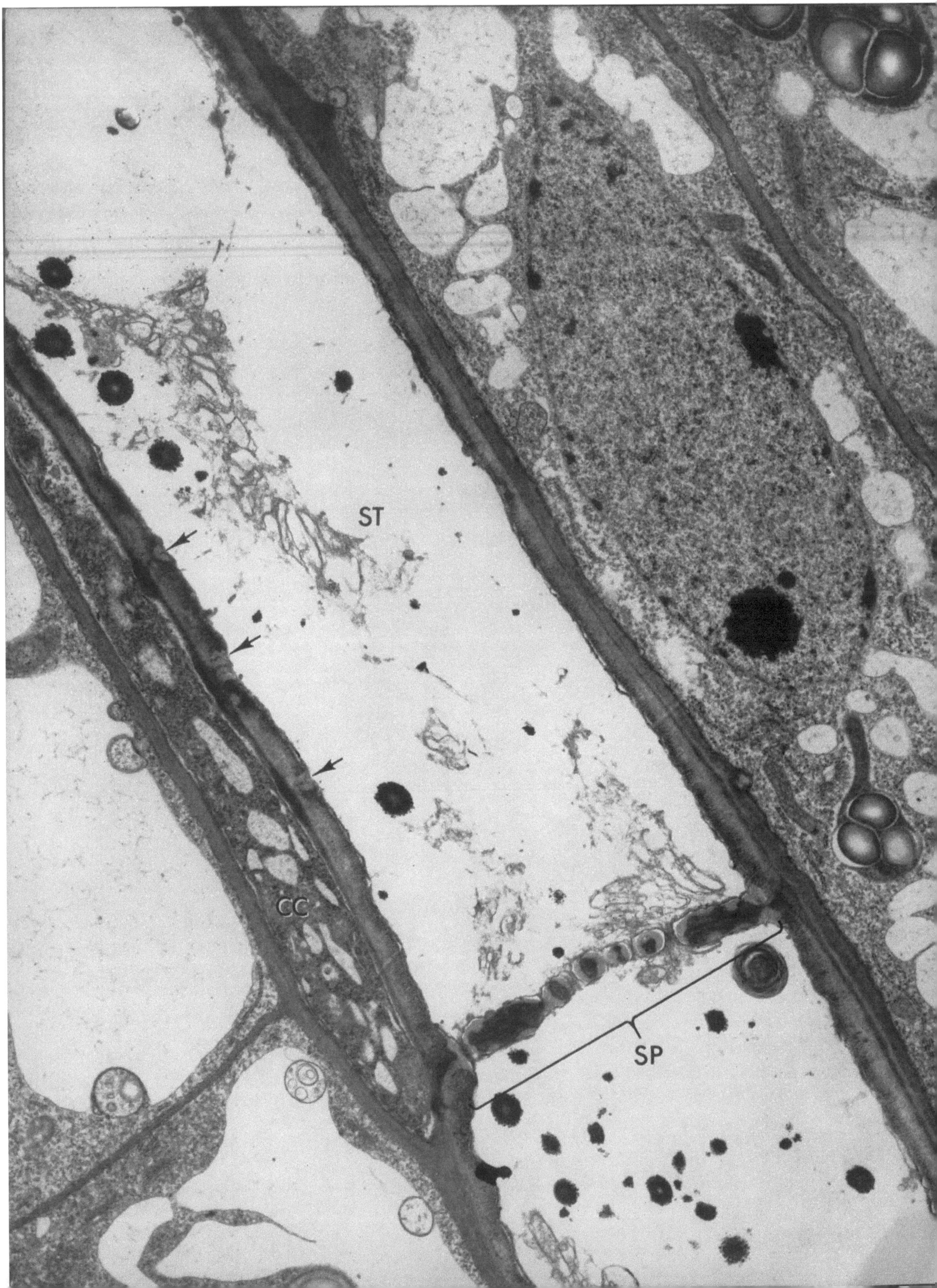
ST
CC
SP

Plate 5.3

Primary Phloem

Sieve elements, the major components of the phloem, contrast sharply in their structure with that of the vascular elements of the xylem. The walls, primary in nature (except in coniferous species), are much thinner; they retain a residual cytoplasm throughout their life as active vascular elements; individual elements connect with one another through sieve areas, which are clusters of very narrow pores; and they occur in close association laterally not with other sieve tubes but with living parenchymal units called companion cells. This tissue functions in the transport of organic nutrients (sucrose at concentrations as high as 30%) from sites of photosynthesis and storage to regions of growth, where structural and nonstructural components of the new tissue are being synthesized.

This micrograph of phloem elements from the stem of *Arabidopsis* is included to give some impression of the characteristics of this important vascular tissue. The two sieve elements, which occupy the center of the image, are relatively narrow thin-walled cells (ST). They have reached a stage of maturity where they are thought to be engaged in active transport of sucrose. The two units are separated by a transverse wall, which includes a sieve area and which is called a sieve plate (SP). It is difficult to tell whether all of the pores are patent, because some are narrower than the section is thin, but it is probable that most of them are. Modifications in the density of the wall around the pores indicates that callose (the material of lower density) formation is in progress and that some of the pores are almost occluded.

Transformation to the mature state involves striking changes in these cells. The nucleus disintegrates and only a few cytoplasmic organelles including plastids and mitochondria remain. Finally even these disappear and scattered vesicles and membranes alone persist as reminders of the former cytoplast. It is characteristic of these sieve elements (ST) to show multiple invaginations of the plasma membrane as the cells age (Plate 5.3.1). Though these may become striking as the tube matures into a transport mechanism, no evidence for their involvement in this process has thus far been presented.

The adjacent cells in this micrograph, except for one, are taken to represent parenchymal elements of the phloem tissue. As such they may be available to replace, in the sense of substituting for, damaged sieve elements. There is nothing in their appearance here to suggest that they are more than relatively undifferentiated, essentially meristematic cells. The single exception to these comments is the small cell to the left of the sieve tube, represented here by only a slice off a narrow margin of the larger unit. This is the companion cell (CC), found usually in close physical relation to sieve elements. It has a cytoplasm rich in ribosomes and mitochondria, plastids without starch, and a prominent nucleus (Plate 5.3.1). Plasmodesmata are common (and morphologically complicated) in the primary walls between this cell and the sieve elements (arrows). Here in this image they are very slender and surrounded by a zone of low density resembling callose. The function of these and, more broadly, the whole companion cell is a subject of intense speculation. There are, however, reasons to suggest that they participate in the translocation going on in the associated sieve tube.

The sieve plate (SP), which is shown to better advantage in text Fig. 5.3a, starts out as typical primary cross wall (with plasmodesmata) between two sieve elements. As these differentiate, the cross wall is remolded. Certain of the plasmodesmata increase in diameter to become pores, and these come to be lined with callose. Eventually the pores fill with "slime", which has a filamentous component, a phloem protein (P. protein), possessing unique structural features. If, as is supposed, sieve tubes in this phase of differentiation are still transporting sucrose against relatively high back pressures, it is reasonable to view the thickness and apparent sturdiness of the sieve plate as important structurally in this phenomenon. In the absence of other obvious candidates for the role of pump, the pore in the plate with its unusual fibrous component emerges as a possible transducer. Furthermore, the whole sieve plate complex possesses the properties of a valve, for evidence is available that the callose can close the plate in a matter of seconds if the sieve tube is damaged.

There are, concerning sieve tubes and their associated companion cells, a number of conflicting interpretations and theories. The nature of the callose, its influence on transport, the role of the slime body and the relation of the companion cell to the functioning sieve element are topics extensively discussed. The review of these topics in this context seems unwarranted, but the reader is advised to consult the original papers for their discussion.

From the stem of *Arabidopsis thaliana* L.
Magnification × 10,500

5.3 Supplemental Reading

Cronshaw, J., Esau, K.: P-protein in the phloem of *Cucurbita*. I. The development of P-protein bodies. J. Cell Biol. **38**, 25–39 (1968).

— — Tubular and fibrillar components of mature and differentiating sieve elements. J. Cell Biol. **34**, 801–815 (1967).

Currier, H. B., Shih, C. Y.: Sieve tubes and callose in *Elodea* leaves. Amer. J. Botany **55**, 145–152 (1968).

Engleman, E. M.: Sieve element of *Impatients sultanii*. I. Wound reaction. Ann. Botany (Lond.) **29**, 83–100 (1965).

Esau, K.: The phloem. Berlin: Gebrüder Borntraeger 1969.

— Cronshaw, J.: Plastids and mitochondria in the phloem of *Cucurbita*. Canad. J. Botany **46**, 877–880 (1968).

Eschrich, W.: Beziehungen zwischen dem Auftreten von Callose und der Feinstruktur des primären Phloems bei *Cucurbita ficifolia*. Planta (Berl.) **59**, 243–261 (1963).

Evert, R. F., Derr, W. F.: Callose substance in sieve elements. Amer. J. Botany **51**, 552–559 (1964).

Kollmann, R., Schumacher, W.: Über die Feinstruktur des Phloems von *Metasequoia glyptostroboides* und seine jahreszeitlichen Veränderungen. II. Mitteilung: Vergleichende Untersuchungen der plasmatischen Verbindungsbrücken in Phloemparenchymzellen und Siebzellen. Planta (Berl.) **58**, 366–386 (1962).

Maxe, M.: Etude de la degenerescence nucleaire dans les cellules criblees de "Polypodium vulgare" (Polypodiacee). C. R. Soc. Biol. (Paris) **262**, 2211–2214 (1966).

Northcote, D. H., Wooding, F. B. P.: Development of sieve tubes in *Acer pseudoplatanus*. Proc. roy Soc. B **163**, 524–537 (1966).

Tamulevich, S. R., Evert, R. F.: Aspects of sieve element ultrastructure in *Primula obconica*. Planta (Berl.) **69**, 319–337 (1966).

Worley, J. F.: Injection of foreign particles and their intracellular translocation in living phloem fibers. Planta (Berl.) **68**, 286–291 (1966).

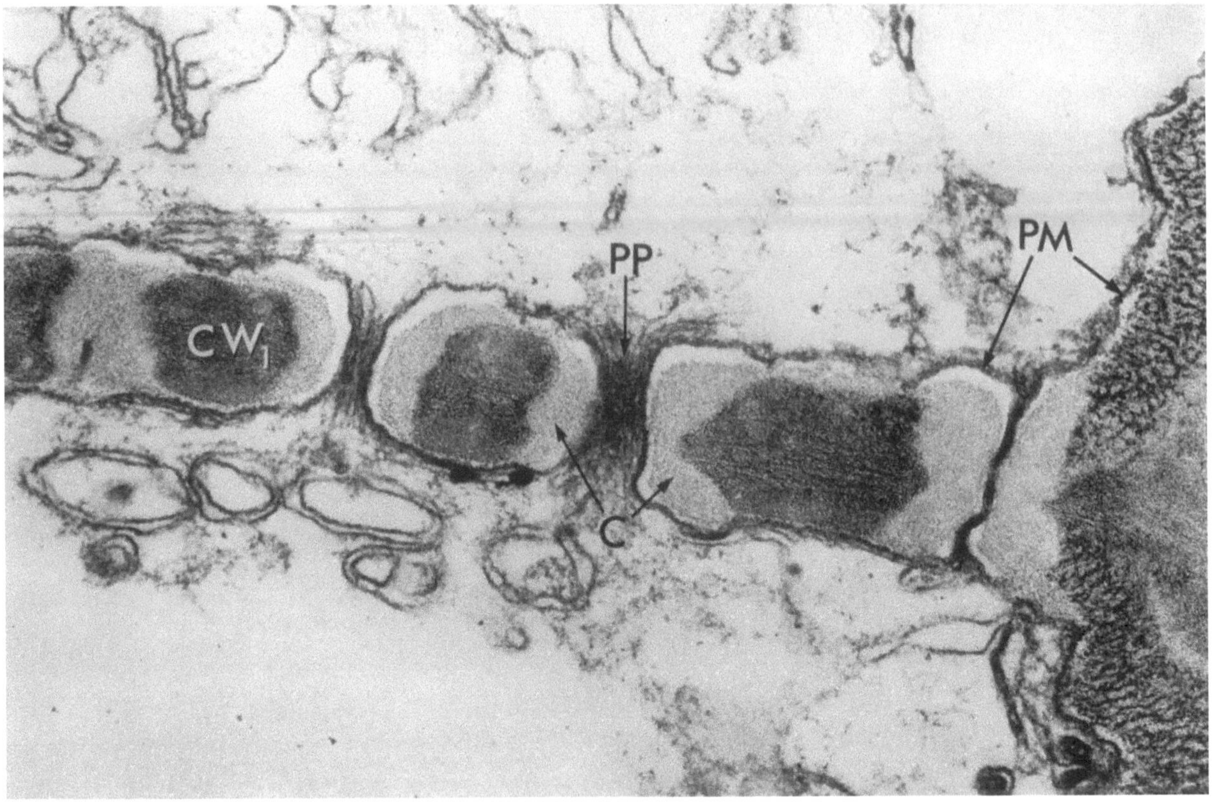
PP
PM
CW1
C

Text Fig. 5.3a

Sieve Plate

This shows at an appropriate magnification the structure of a sieve plate as found in a sieve tube of *Arabidopsis*. At least two pores are transected by the section. They are each about 100 nm in diameter. The plasma membrane (PM) of the contiguous sieve elements are continuous through and around the pores. Callose (C) just within the membrane seems to replace the more dense material of the primary wall (CW_1). The pore space is occupied by numerous 15 nm thick filaments of P-protein (PP), which in higher resolution micrographs are observed to be striated. Remnants of the cytoplasm of the erstwhile living sieve element are evident as membranes on both sides of the plate.

From stem tissue of *Arabidopsis thaliana* L.
Magnification $\times$ 46,000

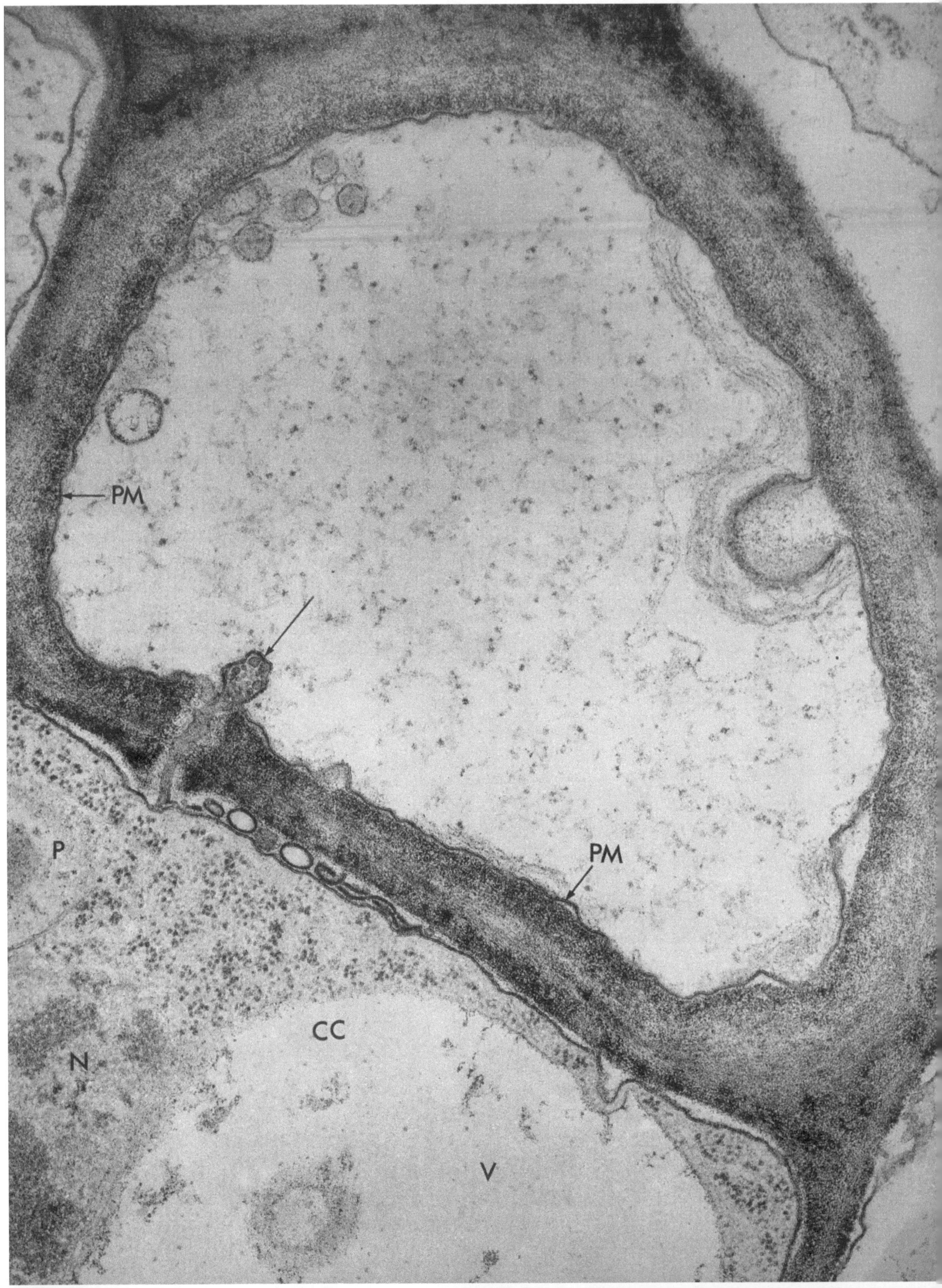
PM
PM
P
CC
N
V

Plate 5.3.1

Sieve Element and Companion Cell

The sieve tube-companion cell relationship seems, as mentioned previously, a difficult one to define. It has long been assumed that the companion cell accepts the role of monitor of the sieve element after the latter's nucleus breaks down and its function is narrowed to that of transport. But what is implied by the word "monitor" or by what means the companion cell performs any such role remains obscure.

This micrograph depicts in cross section a relatively thick-walled sieve element and an adjacent companion cell (CC). The cytoplasm of the sieve element is characteristically free of several organelles but is still limited by an intact plasma membrane (PM). Fine fibrous material on the right hand side of the cytoplast could represent a component of the slime body. The thinner walled companion cell (CC) is by comparison an undifferentiated unit. It possesses a nucleus (N), a vacuole (V) with intact tonoplast, plastids (P) and ribosome-rich cytoplasm.

The structure of greatest interest here is the single pore caught in the plane of section (arrow). It obviously represents a "continuity" between companion cell and sieve tube element. Actually the image is difficult to interpret. The best that can be made of it suggests that the companion cell sends a slender extension through the lumen of the pore to the extracellular space just outside the PM of the sieve element to coincide with a small invagination of the sieve element at this point (arrow). The plasma membrane around the invagination or vesicle apparently belongs to the sieve tube element. Whether this structure represents a special device for the concentration and transport of sucrose from sieve element to companion cell or a mechanism for exchange of energy rich metabolites in the opposite direction may be determined by future studies.

From the protophloem of a leaf of *Phleum pratense* L.
Magnification $\times$ 78,000

5.3.1 Supplemental Reading

Bouck, G. B., Cronshaw, J.: The fine structure of differentiating sieve tube elements. J. Cell Biol. **25**, 79–95 (1965).

Esau, K., Cronshaw, J., Hoefert, L. L.: Relation of beet yellows virus to the phloem and to movement in the sieve tube. J. Cell Biol. **32**, 71–87 (1967).

Gunning, B. E. S., Pate, J. S., Briarty, L. G.: Specialized "transfer cells" in minor veins of leaves and their possible significance in phloem translocation. J. Cell Biol. **37**, C 7–C 12 (1968).

Wooding, F. B. P., Northcote, D. H.: The fine structure and development of the companion cell of the phloem of *Acer pseudoplatanus*. J. Cell Biol. **24**, 117–128 (1965).

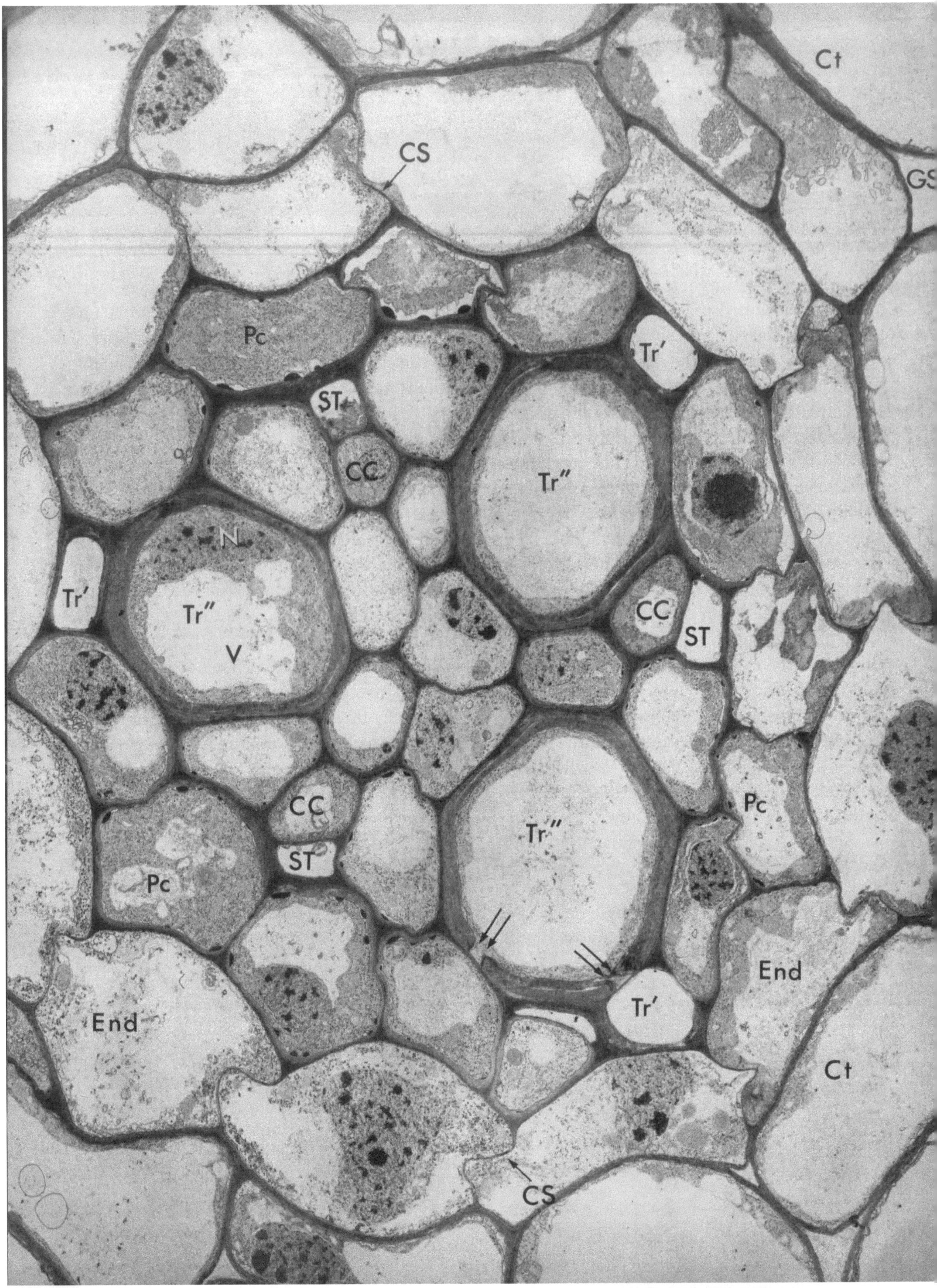
Ct
GS
CS
Pc
Tr′
ST
CC
Tr″
N
Tr′
Tr″
V
CC
ST
Pc
CC
ST
Pc
Tr″
End
Tr′
End
Ct
CS

Plate 5.4

Vascular Cylinder from Root

The primary root possesses a single, centrally located cylinder of vascular tissue within which one can readily identify tracheids, sieve tubes, companion cells, etc. all separated from the larger cortical cells by a layer of special units making up the endodermis. This whole vascular tissue begins to function early in transporting organic nutrients from the seed to the growing tip of the root and, slightly later, water and inorganic nutrients to the growing shoot. It is a tissue possessing several unique and interesting properties including an organization based on the disposition and division patterns of a group of initials in the meristem of the growing root.

One such vascular cylinder is illustrated in this micrograph as it appears in cross section. Though incomplete, the differentiation of its component cells has progressed far enough to provide the observer with distinguishing characteristics. Thus the prominent thick-walled elements, three in number near the center of the bundle, are tracheids or vessel elements (Tr''). Each of these still contains a viable protoplast with nucleus (N) and vacuole (V), and in this respect they are less mature than the smaller, thick-walled elements (Tr') placed just peripherally in the xylem of this primary vascular cylinder. These smaller xylem elements make up the protoxylem, the larger ones the metaxylem. Good examples of simple pits through the thick walls of the larger tracheids are illustrated at the double arrows.

The phloem elements of this vascular tissue are distributed precisely at midpoints between the xylem. Though relatively inconspicuous in size, the sieve elements (ST) can be identified by their thin primary walls and the small invaginations of their plasma membranes. Cells of a similar size, closely adjacent and separated from the phloem elements by an especially thin wall, are interpreted as companion cells (CC). Less differentiated units, some with thin secondary walls, fill up the central parts of the cylinder between the more fully differentiated vascular cells. A layer of living cells of nearly uniform size, called the pericycle (Pc), completes the complement of units in the vascular bundle and in fact marks the outer limits of this tissue. In branch roots, which develop secondarily, the pericycle provides the initials for the meristem from which the new vascular tissue will take its origins. These cells of the pericycle show a thin secondary wall and between it and the plasma membrane numerous dense deposits. The latter are thought to represent suberin, a substance that is possibly important in controlling water exchange between the vascular elements and surrounding cortical tissues.

Finally, there is, within this field yet another tissue possessing unique properties. It is the layer of cells just outside the pericycle and commonly referred to as the endodermis (End). Its morphogenetic origins as well as other features relate it to the cortical tissues of the root rather than to the vascular cylinder. Thus it is more properly thought of as the innermost layer of the cortex, but one that is highly specialized for controlling the ionic environment within the vascular tissue. This it does in spite of large fluctuations in these aspects of the environment external to the root (as encountered, for example, in brackish water).

Endodermal cells (End) are fairly uniform in size and structure. They are unique in showing very thin walls along the sides adjacent to other endodermal cells in the cylinder, and it is within these walls that one can identify the so called Casparian strip (CS), pictured at higher resolutions and discussed in greater detail in the next plate (5.4.1). The slightly folded contour of this wall is probably a product of preparation procedures.

Parenchymal cells with huge vacuoles but otherwise of undistinguished character make up the cortex (Ct) just peripheral to the endodermis. These cells, which do not divide, may differentiate in some species into schlerids or fibers, or may develop banded thickenings. The cortex persists in monocotyledons, such as the timothy shown here, but is sloughed off in roots which display secondary growth.

From a small growing root of timothy, *Phleum pratense* L.

Magnification × 5,200

5.4 Supplemental Reading

Esau, K.: Plant anatomy, 2nd ed. New York: John Wiley & Sons 1965.

Torrey, J. G.: Development in flowering plants. New York: Macmillan Co. 1967.

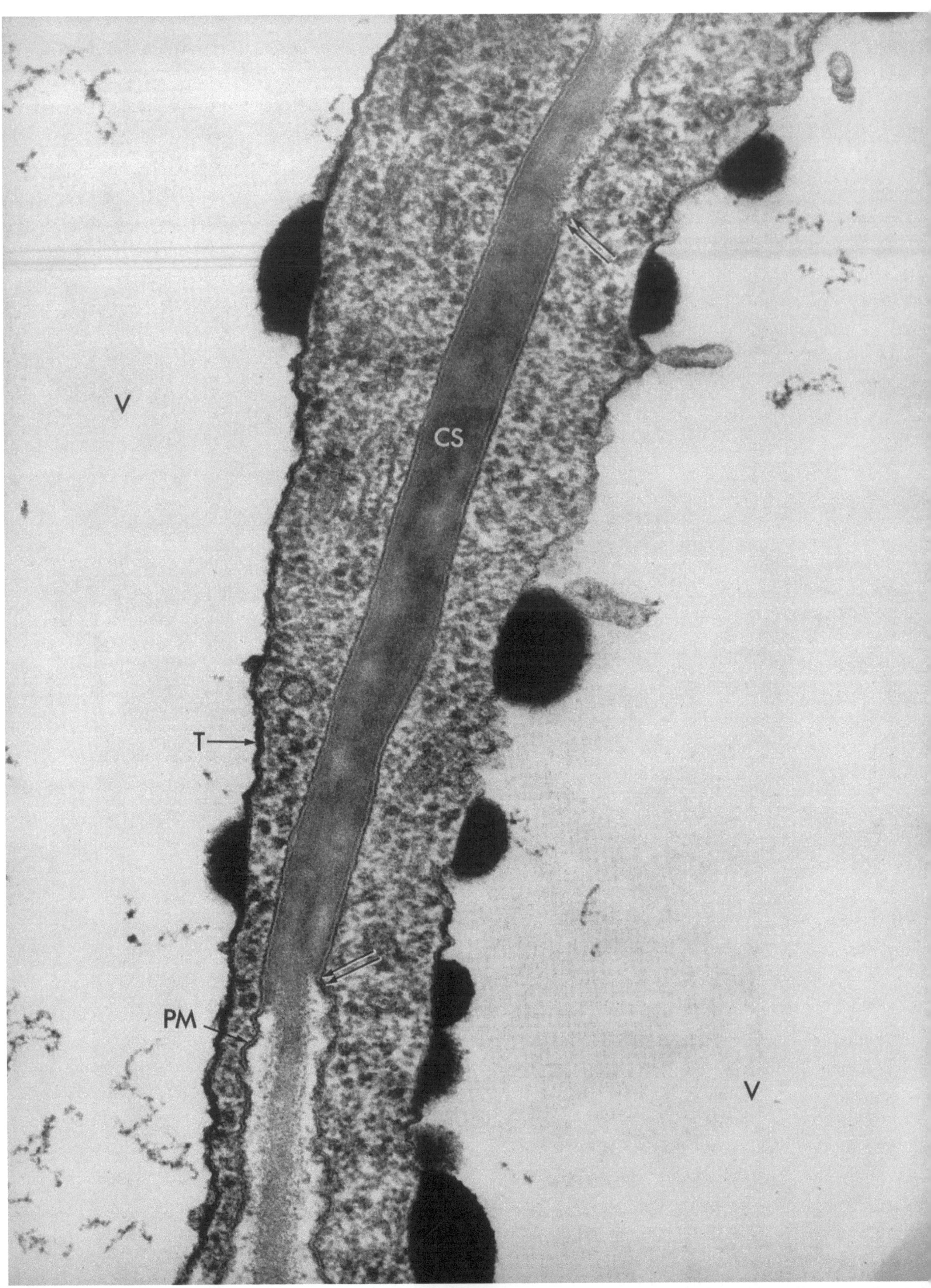

V
CS
T
PM
V

Plate 5.4.1

Casparian Strip

The development of an endodermis around a vascular cylinder is an almost universal feature of roots. In stems it is often replaced by a starch sheath, a layer of cells in which starch-containing plastids are prominent. But starch is lacking in endodermal cells proper both in roots and stems. When an endodermis develops in light-grown green stems, it is isolated from the chlorophyllous cells by a layer of cells lacking chlorophyll. It has been proposed that antioxidants produced by chlorophyllous cells inhibit the oxidation of precursors of the Casparian strip.

This interesting structure is the distinguishing feature of the endodermis. It is a band of lignified or suberized material which circumscribes the cell along its anticlinal walls (text Fig. 5.4.1a). In this micrograph the strip (Cs) is seen in the transverse section of a radial wall separating two endodermal cells. This peculiar wall (the strip) is distinguished by its even texture, smooth surface and a higher density along its surface where it is continuous with the outer leaflet of the plasma membrane (PM). It contrasts strikingly with the relatively loose fibrillar nature of the adjacent wall material.

It is noteworthy that the plasma membrane is so closely adherent to the wall throughout the width of the strip. This is unusual among plant cells and is probably related to the role of the strip in isolating the vascular cylinder from its environment. The degree of bonding in this contact between the protoplast and the strip has been demonstrated by experiments on plasmolysis. The protoplast remains attached to the strip even after it has been pulled away from all other points along the wall. In this respect and in its resistance to the passage of diffusible molecules, this structure is analogous to the "tight junction" found in some animal tissues. Except for the barrier provided by the Casparian strip, ions from the root environment could enjoy unimpeded diffusion along the primary walls of the parenchymal cells and thence enter the xylem stream. However, the Casparian strip assures that entering ions must pass through the plasma membranes of the endodermal protoplasts where selective mechanisms may determine the composition of the solution to be transported to the upper parts of the plant. Furthermore, there are no air spaces between endodermal cells, so that the strip constitutes an uninterrupted block to the diffusion of gases between the vascular cylinder and the root cortex. And finally, the lignified strip is so resistant to hydrolytic enzymes that it effectively prohibits the passage of pathogens across the endoderm and into the stele.

Marginal to the Casparian strip (double arrows) the cell wall appears loose and fibrous. The plasma membrane (PM) is characteristically wavy in outline, a feature which is probably related to the radial growth of the endodermal cells in areas adjacent to the strip. At the corners of the endodermal cells there are frequently clusters of microtubules.

The vacuoles (V) are limited by a prominent tonoplast (T). The dense blobs on its inner surface may be tannins condensed out of the vacuole contents.

From the root of *Limonium sinuatum* Mill.
Magnification ×140,000

5.4.1 Supplemental Reading

Arnold, A.: Über den Funktionsmechanismus der Endodermiszellen der Wurzeln. Protoplasma (Wien) **41**, 189–211 (1952).

Bonnett, H. T., Jr.: The root endodermis: fine structure and function. J. Cell Biol. **37**, 109–205 (1968).

Falk, H., Sitte, P.: Untersuchungen am Caspary-Streifen. In: Proc. European Regional Conf. Electron Microscopy, Delft, vol. 2, p. 1063 (Houwink, A. L., and Spit, B. J., eds.). Delft: De Nederlandse Vereniging voor Electronenmicroscopie 1961.

Fleet, D. S. van: Histochemistry and function of the endodermis. Botan. Rev. **27**, 165–220 (1961).

Ledbetter, M. C.: The Casparian strip: a site of a functional tight junction in plant roots. J. Cell Biol. **35**, 79A (1967).

Scholander, P. F., Hammel, H. T., Bradstreet, E. D., Hemmingsen, E. A.: Sap pressure in vascular plants. Science **148**, 339–346 (1965).

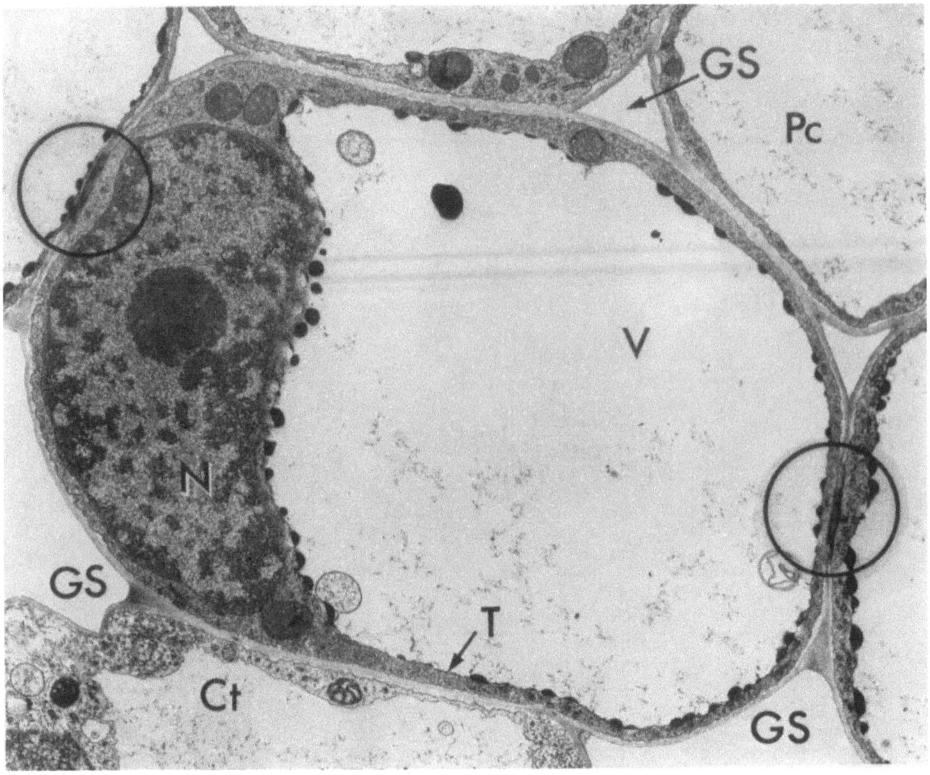
GS
Pc
V
N
GS
T
Ct
GS

Text Fig. 5.4.1a

Endodermal Cell

This shows a single endodermal cell with Casparian strip in section on opposite anti-clinical surfaces (see circles). The cell contains a prominent nucleus (N) and a large vacuole (V) limited by the tonoplast (T). Cells of the pericycle (Pc) are at the top of the micrograph and those of the cortex (Ct) at the bottom. Gas spaces (GS) appear between the cells on both sides of the endodermis but never interrupt the Casparian strip.

From root tissue of *Limonium sinuatum* Mill.

Magnification $\times$ 8,600

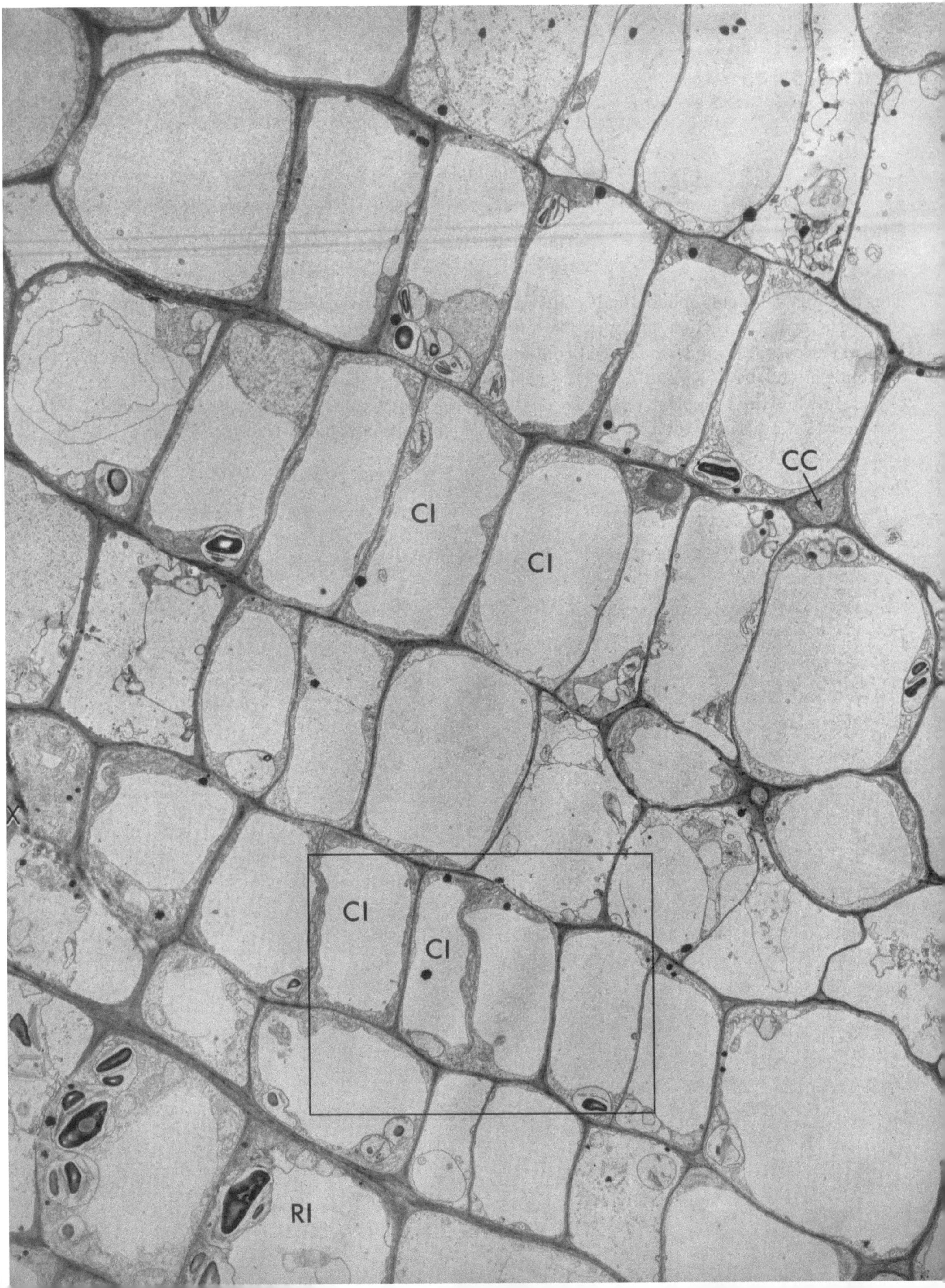
CC
CI
CI
CI
CI
RI

Plate 5.5

Vascular Cambium

The vascular cambium exists as a cylinder of meristematic cells within woody stems. From its inner side it contributes new cells to the secondary xylem and from its outer surface to the secondary phloem. It remains active throughout the life of the plant.

In this micrograph the zone of the cambium observed in a cross section of the stem, runs approximately from lower left to upper right. Although their certain identification is difficult, the cambial initials are probably those cells marked CI. These are labeled as such on the basis of their location and because they lie on opposite sides of a newly formed cell plate. These criteria are not, however, entirely reliable since it is known that progeny of the initials destined to become xylem or phloem elements may divide after their determination and before beginning their differentiation.

Among meristematic cells, these cambial initials are remarkable in having very large vacuoles and in any one section seemingly little ribosome-rich cytoplasm. The cross sectional profile of these cells shown here is misleading when one tries to imagine the three-dimensional form of the whole unit. Actually initials such as these are long fusiform cells (170 μm long and only 7 μm wide or thick). The volume of cytoplasm may be as great as in any isometric meristematic cell, but it is spread out into a thin layer between the tonoplast (vacuole membrane) and the plasma membrane. The cytoplasm, which is continuous over the entire cell, contains a few small plastids (some with starch grains), mitochondria, dictyosomes, and ribosomes, all of which lack any unusual features (Plate 5.5.1). In spite of their highly anisometric shape and the huge vacuoles, these cells are able to form a phragmoplast and a new cell plate parallel to their long axis and periclinal to the cambial cylinder.

This process of proliferation contributes new cells to the files of tracheary elements and sieve elements which run from left to right and slightly diagonally across this micrograph. The xylem elements, can be recognized on the left by their thickened walls. Within the area of the micrograph they retain their living protoplasts, which are of course engaged mostly in wall formation. The phloem elements to the right are distinguished by their thinner walls and the appearance of other features already reviewed (Plate 5.3). A few possible companion cells (CC) are included in the field. All of these derivatives are, like the initials, relatively long, spindle-shaped cells.

This micrograph is valuable in showing two additional features of this remarkable tissue. It is obvious and well known that, as the cylinder of secondary xylem increases in diameter, its circumference also increases and that more cambial initials are included circumferentially in the cambial cylinder. It follows that not all divisions in the cambium can be periclinal, some must be radial or anticlinal. The product of one such division is included at the left (at X), where a single file of xylem elements abruptly becomes a double file and procedes in this doubleness outward through cambium and phloem to the right.

In addition, the micrograph at the lower left includes a few ray cells (RI), identified as such by a thicker layer of cytoplasm around a relatively smaller vacuole. These cells are, of course, more nearly isometric than the cambial initials (CI) described above. They also show a greater tendency to store starch and also presumably to make the products of starch degradation available to the differentiating, wall-producing tracheids near by. A ray cell initial (RI) within the cambium is probably included at bottom center. It has a dense, ribosome-rich cytoplasm characteristic of rapidly proliferating cells.

The rectangular area outlined is shown at higher magnifications in Plate 5.5.1.

From the woody stem of common locust, *Robinia pseudoacacia* L.

Magnification × 5,700

5.5 Supplemental Reading

Bailey, I. W.: The cambium and its derivative tissues. III. A reconnaissance of cytological phenomena in the cambium. Amer. J. Botany **7**, 417–434 (1920).

Morey, P. R., Cronshaw, J.: Induced structural changes in cambial derivaties of *Ulmus americana*. Protoplasma (Wien) **62**, 76–85 (1966).

Srivastava, L. M., O'Brien, T. P.: On the ultrastructure of cambium and its vascular derivatives. I. Cambium of *Pinus strobus* L. Protoplasma (Wien) **61**, 257–276 (1966).

Srivastava, R. H., O'Brien, T. P.: On the ultrastructure of cambium and its vascular derivatives. II. Secondary phloem of *Pinus strobus* L. Protoplasma (Wien) **61**, 277–293 (1966).

Wetmore, R. H., DeMaggio, A. E., Rier, J. P.: Contemporary outlook on the differentiation of vascular tissues. Phytomorphology **14**, 203–217 (1964).

Wilson, B. F.: Mitotic activity in the cambial zone of *Pinus strobus*. Amer. J. Botany **53**, 364–372 (1966).

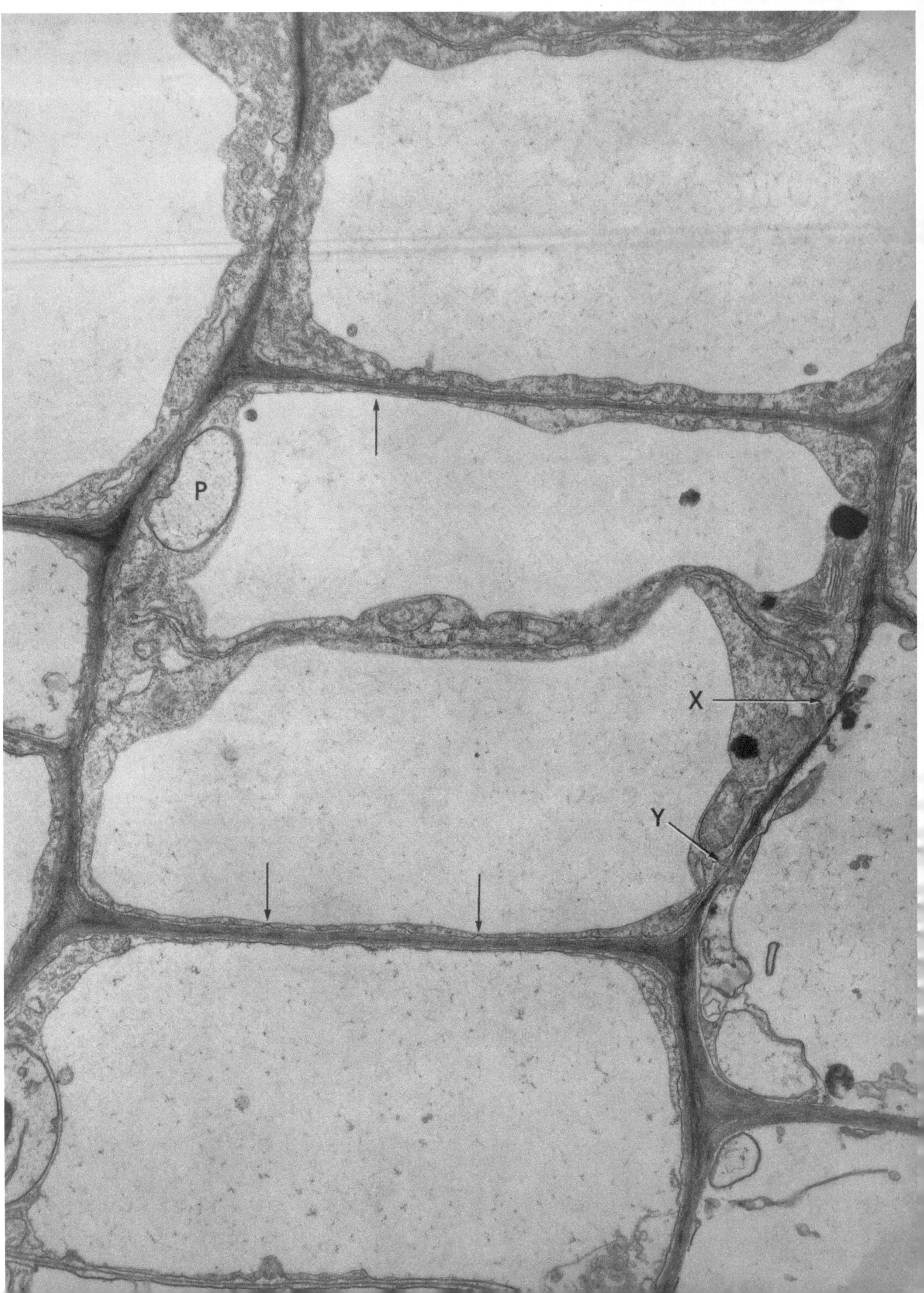
P
X
Y

Plate 5.5.1

Cambial Initials

The meristematic cells of the cambium, called cambial initials, are so unusual, especially among meristematic cells, that they deserve a brief discussion of their own. Unlike most other meristematic cells they possess a large vacuole and are fusiform in shape, being 20–30 times as long as they are wide or thick. This means that in periclinal division a prominent tonoplast must divide and that the phragmoplast, which begins at the center of the mitotic spindle, must spread progressively toward the extreme tips of the cell before cytokinesis is complete.

Some of these features and problems can be appreciated from an examination of the image in this Plate. It includes at a much higher magnification (and a 90° rotation) a few of the cells outlined in Plate 5.5. A file of cells crosses the picture from lower left to upper right. At its lower end there is shown a part of a young phloem cell (sieve tube element) and at its upper end a new xylem element. The cells in the center are newly formed. In fact, at its left end the plate separating them has not quite reached the plasma membrane of the parent cell. The new zone of separation, i.e., the new plate, is extremely thin and shows only the beginnings of a middle lamella. There is no clear indication which of the two daughter cells will remain the initial and which will become either a sieve element or a xylem element.

The thin layers of cytoplasm that surround the vacuoles display the usual organelles and systems. Mitochondria and proplastids (P) are readily identified. There are also vesicles of the ER, ribosomes, and small stacks of flattened vesicles, the dictyosomes. In places, especially against the tangential walls, the cytoplasm becomes so attenuated that the inner surfaces of the plasma and tonoplast membranes are in contact (see arrows).

In their tangential dimension (left to right dimension on picture) the initials as well as the new xylem and phloem cells remain constant during subsequent growth. But in their radial dimension they increase from the time they are established until they are mature units enclosed in a thickened, cellulose wall. This increase is accomplished through a temporary softening of the primary wall, first of the mother cell and second of the xylem or phloem initial itself. One such thinning and "depolymerization" of the side wall is seen at X, where the plate meets the wall. This may coincide with a region where plasmodesmata are numerous. Another is seen at Y, where the radial walls of the initials are softening at another point. It is interesting that as this process proceeds, the plasma membranes of the protoplasts immediately contiguous become indistinct. Whether this results from activity involving the secretion of a depolymerase (cellulase) or whether it is descriptive of an early stage in wall formation is not at once evident. This question and others concerning the division and growth of cambial initials are ripe for further investigation.

From the woody stem of common locust, *Robinia pseudoacacia* L.
Magnification × 23,000

5.5.1 Supplemental Reading

Bailey, I. W.: The formation of the cell plate in the cambium of the higher plants. Proc. nat. Acad. Sci. (Wash.) **6**, 197–200 (1920).

Cronshaw, J.: The organization of cytoplasmic components during the phase of cell wall thickening in differentiating cambial derivatives of *Acer rubrum*. Canad. J. Botany **43**, 1401–1408 (1965).

Goldstein, B.: A study of progressive cell plate formation. Bull. Torrey Botan. Club **52**, 197–219 (1925).

Wardrop, A. B.: Cellular differentiation in xylem. In: Cellular ultrastructure of woody plants, p. 61–97 (Cote, W. A., ed.). Syracuse, N.Y.: Syracuse University Press 1965.

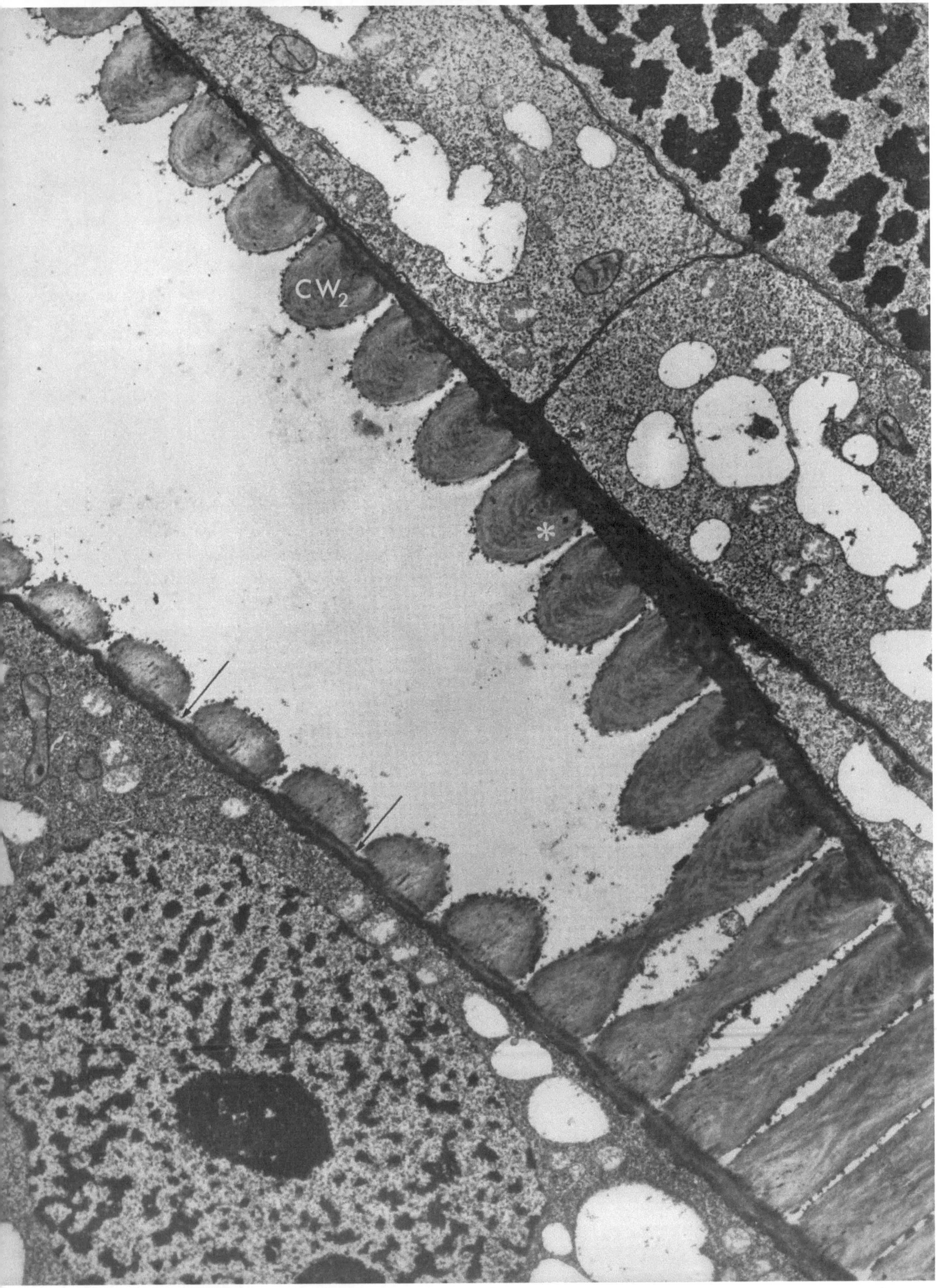

CW_2

Plate 5.6

Primary Xylem

In the development of a vascular bundle within a stem or root, it is common for the xylem elements first formed to exhibit some characteristics regarded as primitive. Among these, the deposition of the secondary wall in either spiral or ring thickenings is especially representative. This repeats in essence the form of the tracheids found in the earliest evolutionary forms of vascular plants, extinct or extant.

Such a primitive xylem element is shown in longitudinal section in this micrograph. The cell runs from lower right to upper left and is enclosed on opposite sides by living parenchymal cells of this tissue. Actually the section with respect to the "tracheid" is medial longitudinal at the upper left and more nearly tangential at the lower right. Thus the secondary thickenings of the wall (CW_2) are cut transversely at one end and obliquely-longitudinally at the other. In places (as depicted) where the cell has not been stretched in length by the growth of adjacent tissues, the secondary thickenings are close together; but where elongation has been forced, the distance between them may exceed by 2 or 3 times the width of the thickening itself.

There is accumulating evidence that differentiation within the cytoplasm of the tracheid initial precedes the localized deposition of secondary wall material. Thus the cytoplasm representing presumptive zones of wall deposition becomes more densely populated with organelles. Microtubules anticipate not only the position of the bands but also the circumferential orientation of the long axis of the thickening whether it be spiral or annular (see Plate 2.4). Cytoplasmic streaming has been observed to be more active in these zones. The wall is obviously deposited in phases, as denoted by the layering in its sectioned image (see *). In other views it is evident that the microfibrils of cellulose run parallel to the orientation of the cytoplasmic microtubules. This organization of wall subunits is thought to impart extraordinary tensile strength to these bands in the direction of their length. A further strengthening of the secondary wall is achieved by lignification, an impregnation that acts to bond the microfibrils and prevent slippage of one past the other.

As these early xylem elements mature and elongate, the primary walls at their ends and between the circular or helical bands of secondary wall are attacked by enzymes and may be wholly or partially degraded. This is accompanied by or followed by autolysis of the cytoplast, the remnants of which are alone evident in this micrograph. The removal of the primary walls between the rings (see arrows) provides for free access to and passage of water and solutes between tracheid or vessel and parenchymal cells, especially in leaves. The degradation of end walls opens up the tracheal system to the free movement of the transpiration stream.

From[1] vascular tissue in leaf of *Triticum aestivum* L.

Magnification ×26,000

[1] This micrograph was kindly provided by B. E. S. Gunning.

5.6 Supplemental Reading

Bierhorst, D. W.: Observations on tracheary elements. Phytomorphology **10**, 249–305 (1960).

Cronshaw, J., Bouck, G. B.: The fine structure of differentiating xylem elements. J. Cell Biol. **24**, 415–431 (1965).

Hepler, P. K., Newcomb, E. H.: The fine structure of young tracheary xylem elements arising by redifferentiation of parenchyma in wounded *Coleus* stem. J. exp. Botany **14**, 496–503 (1963).

Northcote, D. H.: Changes in the cell walls of plants during differentiation. Symp. Soc. exp. Biol. **17**, 157–174 (1963).

Pickett-Heaps, J. D.: Xylem wall deposition. Radioautographic investigations using lignin precursors. Protoplasma (Wien) **65**, 181–205 (1968).

Sinnott, E. W., Bloch, R.: The cytoplasmic basis of intercellular patterns in vascular differentiation. Amer. J. Botany **32**, 151–156 (1945).

Wooding, F. B. P.: Radioautographic and chemical studies of incorporation into sycamore vascular tissue walls. J. Cell Sci. **3**, 71–80 (1968).

— Northcote, D. H.: The development of the secondary wall of the xylem in *Acer pseudoplatanus*. J. Cell Biol. **23**, 327–337 (1964).

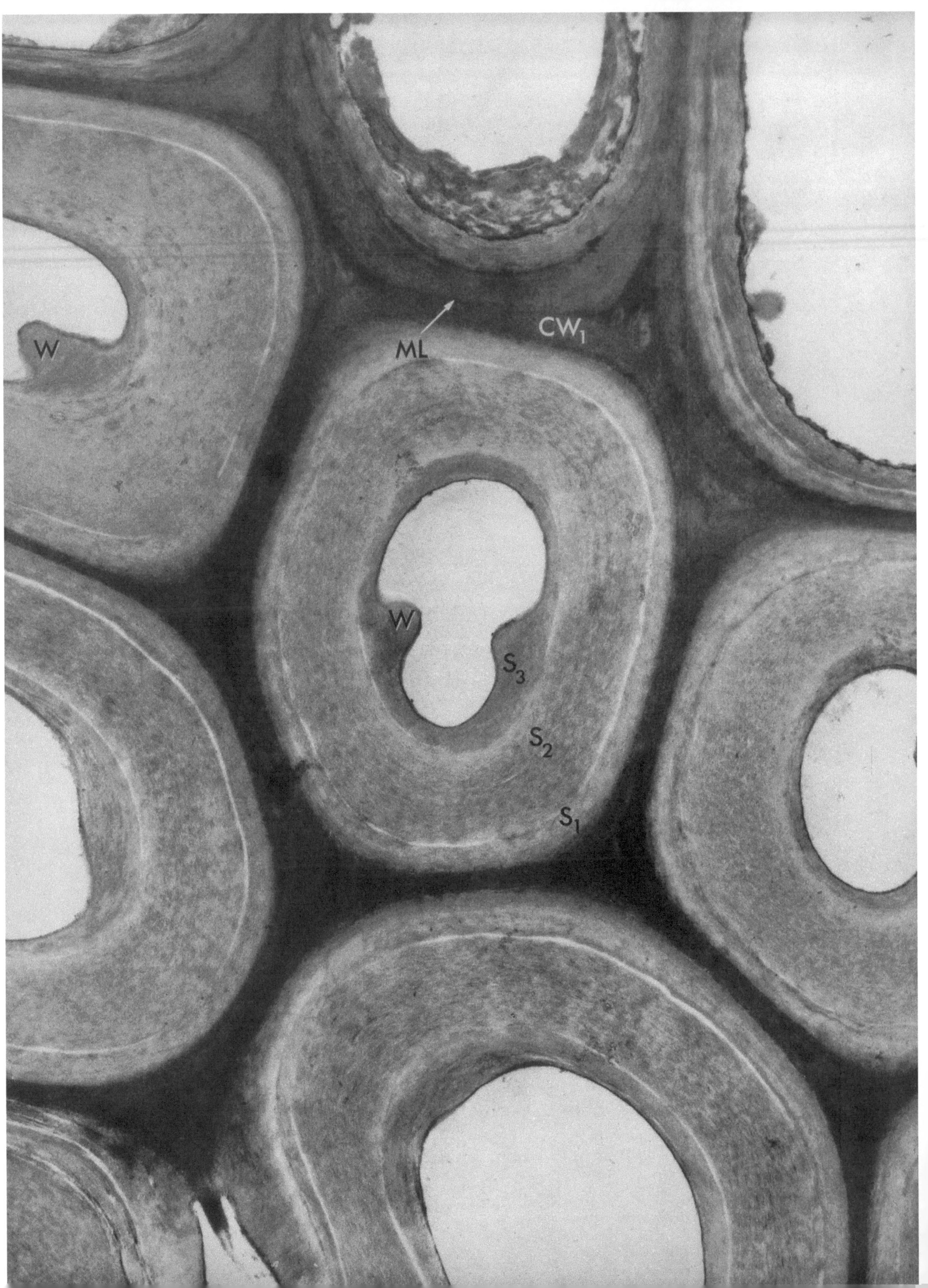
W
ML
CW1
W
S3
S2
S1

Plate 6.1

Fibers

Among plant cells with thick walls, the fibers such as those depicted here are outstanding. Their specialization is a thickened and lignified wall, which gives them tensile strength and rigidity and makes them valuable as supportive elements in the plant. In some examples of these cells, as in this micrograph, the protoplasts are no longer present, having given over almost their entire substance to wall production. In others, however, the protoplasts persist, apparently receiving life-supporting metabolites from cells of adjacent tissues.

Cross sections of fibers are shown in this plate. Parts of seven fibers, included in the lower-half of the micrograph, are readily recognized as being different from the three cells at the top. The latter are probably gelatinous fibers, so called because of the special characteristics of their walls. Each fiber of the first group (of seven) has an apparently empty central cavity surrounded by a wall of relatively low density, which is three micrometers thick. This is the secondary wall (S_1, S_2, S_3). A dense layer of uniform thickness external to it represents the primary wall (CW_1). The latter is known to be more highly lignified than the secondary wall and therefore less available to hydration. A much thinner dark layer that separates the individual fibers and fills in the corners is the middle lamella (ML).

It is clear that layering in the wall of the fiber is more general than that which distinguishes primary from secondary. There is, for example, striking evidence that layers or plies exist in the secondary wall itself. These had earlier been identified by light microscope studies in which it was recognized further that the angle of birefringence shifted from layer to layer (see Plate 5.2). In all there are, as in tracheids and vessels, three layers (see Plate 5.2), identified from outside to inside as S_1, S_2 and S_3. Each is put down during a separate phase in wall formation, each has a distinctive thickness and in each the orientation of the cellulose or other filaments changes direction. In the outermost layer (S_1), the filaments run circumferentially and in a plane nearly normal to the long axis of the fiber. In the thickest layer (S_2), the pitch shifts about 60° so that they spiral steeply around the fiber; and in the innermost, S_3 layer, the filaments, in this case non-cellulosic, are again nearly circumferential. This layered construction in the wall doubtless contributes to the rigidity and strength of the fibers together with that imparted by their lignification. The small elevations on the surface facing the lumen of the fiber are called warts (W) and probably represent the final product of the protoplasmic machine before it stopped synthesis altogether (Plate 5.2.1). The distinct lines of separation between the three innermost layers, as well as the changes in cellulose filament orientation, imply that the living unit interrupted the production of cellulose precursors and altered its cytoplasmic organization over a brief period of time.

As suggested by its name the fiber is a long spindle-shaped cell. Those depicted here, while only 15–20 μm in diameter, may be a millimeter or more in length. Some plant fibers of a similar nature may be several centimeters in length, and because of this and their tensile strength they have great economic value.

From a young stem of locust, *Taxus canadensis* Marsh.
Magnification × 17,000

6.1 Supplemental Reading

Bailey, I. W.: The walls of plant cells. Publ. Amer. Ass. Advanc. Sci. **14**, 31–43 (1940).

— Kerr, T.: The cambium and its derivative tissues. X. Structure, optical properties and chemical composition of the so-called middle lamella. J. Arnold Arboretum **15**, 327–349 (1934).

— — The visible structure of the secondary wall and its significance in physical and chemical investigations of treacheary cells and fibers. J. Arnold Arboretum **16**, 273–300 (1935).

— — The structural variability of the secondary wall as revealed by "lignin" residues. J. Arnold Arboretum **18**, 261–272 (1937).

Czaninski, Y.: Observations infrastructurales sur les fibres libriformes du xyleme du *Robinia pseudoacacia.* C. R. Soc. Biol. (Paris) **264**, 2754–2756 (1967).

Hiller, C. H., Brown, R. S.: Comparison of dimensions and fibril angles of loblolly pine tracheids formed in wet or dry growing seasons. Amer. J. Botany **54**, 453–460 (1967).

Wardrop, A. B., Harada, H.: The formation and structure of the cell wall in fibres and tracheids. J. exp. Botany **16**, 356–371 (1965).

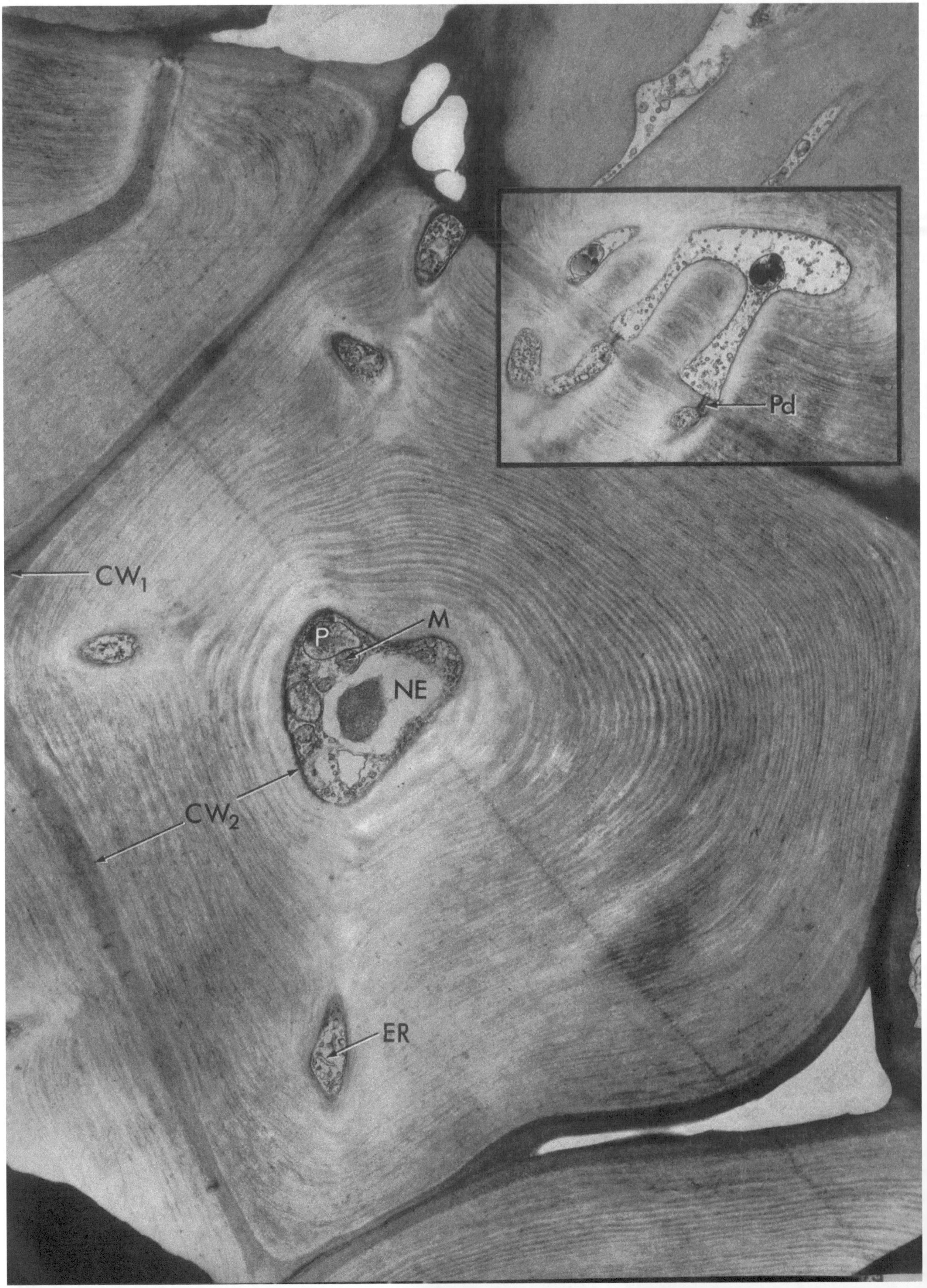
Pd
CW1
P
M
NE
CW2
ER

Plate 6.2

Stone Cells

Among cells that form very thick walls none is more dramatic than the sclereids or stone cells found in the fleshy parts of the fruit of the pear (*Pyrus communis*). These are the little hard bits which impart a gritty texture to this fruit. They occur in small, irregularly branched clusters.

It can be seen in this micrograph that the secondary walls (CW_2) of these cells are multilayered and extremely thick. In spite of this feature the cell retains a living protoplast, which, however, is not confined to a central region. Rather, it sends slender branches in all directions through the secondary walls to the inner surface of the primary wall (CW_1), where they coincide with clusters of plasmodesmata (see insert). There is no evidence of regularity in this branching; indeed it owes its randomness to the morphogenetic history of the cell. Beginning as a sphere with a primary wall (CW_1), the protoplast puts down a secondary wall (CW_2) except in the regions of the plasmodesmata (Pd). With these latter it retains a connection and some physiological continuity with similar extensions of cells in adjacent sclereids. Presumably these connections provide for a uniform distribution of metabolites in a group of protoplasts that could scarcely receive any nutrients or oxygen through such thick walls. They must also permit the continuing uptake of glucose for the cyclic manufacture of more and more wall. The arrangement is reminiscent of that that exists among osteocytes in the Haversian system of calcified bone.

Various organelles and other cytoplasmic components [mitochondria (M), plastids (P), ribosomes, and ER vesicles] can be seen in profile in all parts of the cytoplasm represented in the picture. A small tangential slice off the tip of the nucleus is evident in the center surrounded by the intracisternal space of a dilated nuclear envelope (NE).

From the carpel wall of the pear, *Pyrus communis* L.

Magnification $\times 12{,}000$

6.2 Supplemental Reading

Foster, A. S.: Plant idioblasts: remerkable examples of cell specialization. Protoplasma (Wien) **46**, 184–193 (1956).

Mia, A. J.: Ontogeny and differentiation of sclereids in *Rauwolfia*. Amer. J. Botany **51**, 78–87 (1964).

Sterling, C.: Sclereid development and the texture of Bartlett pears. Food Res. **19**, 433–443 (1954).

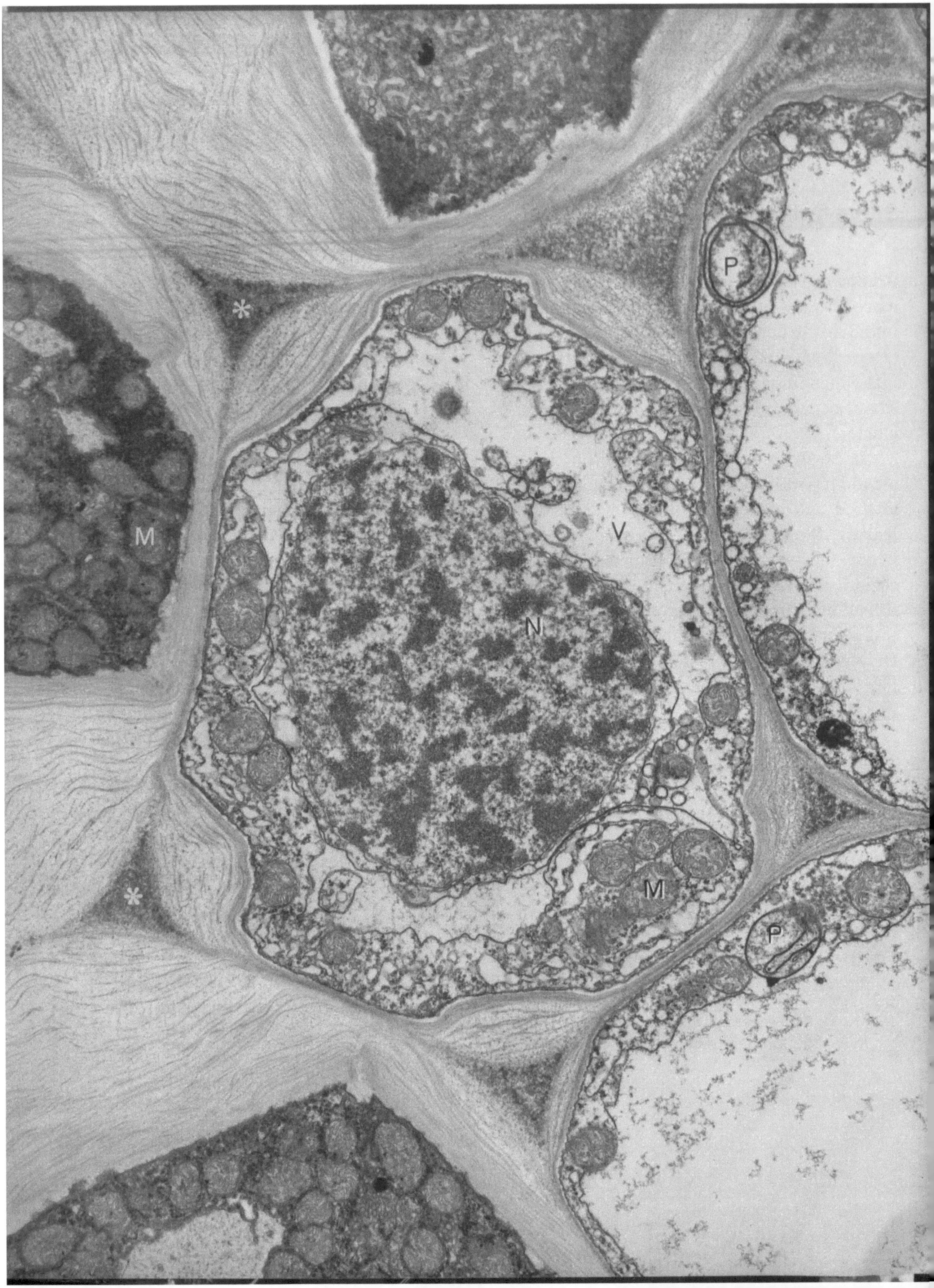
*
M
P
V
N
*
M
P

Plate 6.3

Collenchyma

The most significant differences between the cells of collenchymal and sclerenchymal tissues described in this section (Plates 6.1 and 6.2) are found in the nature of the wall. The thickened walls of collenchymal cells are primary and contain pectin and especially large amounts of water. The thickened portions of the wall may be variously distributed in relation to cell contacts and the plant organ as a whole. The walls are unlignified and are capable of plastic deformation in contrast to the pronounced elastic properties characteristic of lignified fibers. Thus collenchymal tissues are able to provide support for plant parts and at the same time permit further growth through changes in size and shape of individual cells. Collenchymal cells also retain their protoplasts and are usually capable of dividing.

The cells illustrated here are from a cross section of a wheat filament. The thickened corners describe this tissue as representative of angular collenchyma. The intercellular spaces (*) are filled with some dense material, probably pectin. Alternate layers of what appear as particles and fibrils can be seen in the wall. The latter are interpreted as representing microfibrils of cellulose lying within the plain of section and the former as cross-sections of microfibrils oriented more or less normal to the section. The water typical of collenchymal walls must have occupied the abundant open spaces between the microfibrils. Notice that in both orientations the microfibrils lie in planes parallel to the cell surface.

It is evident from the known behavior of wheat filaments that these cells are capable of remarkable and rapid expansion in length. In fact it has been reported that they display the most rapid growth rate known in plants. At the time of opening of the wheat flower the filament may extend at the rate of 2–3 mm per minute. Such growth as this must take place as a result of a rapid change in cell shape accomplished through a plastic extension of the side walls. Energy is doubtless required for this process, and it is notable that portions of cells at the top and at the left in this micrograph are extraordinary in that most of the cytoplasm is occupied by mitochondria (M). Whether microtubules are prominent in these cells during their elongation has not been determined, and the fact that this specimen was fixed in phosphate buffered OsO_4 precludes any demonstration of them here. It may also be significant for their rapid elongation that the cells of these wheat filaments possess prominent tonoplasts and vacuoles (V). Plastids (P) do not develop in these cells beyond the proplastid stage.

From the anther filament of wheat, *Triticum aestivum* L.

Magnification $\times$ 16,500

6.3 Supplemental Reading

Cheignon, M., Schaeverbeke, J.: L'ultrastructure des filets staminaux du *Zea mays* L. et ses modifications sous l'action de l'acide gibbérellique et de l'acide β-indolyl-acétique. C. R. Soc. Biol. (Paris) **260**, 5085–5088 (1965).

Greyson, R. I., Tepfer, S. S.: An analysis of stamen filament growth of *Nigella hispanica*. Amer. J. Botany **53**, 485–490 (1966).

Greyson, R. I., Tepfer, S. S.: Emasculation effects of the stamen filament of *Nigella hispanica* and their partial reversal by gibberellic acid. Amer. J. Botany **54**, 971–976 (1967).

Roland, J.-C.: Infrastructure des membranes du collenchyme. C. R. Soc. Biol. (Paris) **259**, 4331–4334 (1964).

— Edification et infrastructure de la membrane collenchymateuse. Son remaniement lors de la sclerification. C. R. Soc. Biol. (Paris) **260**, 950–953 (1965).

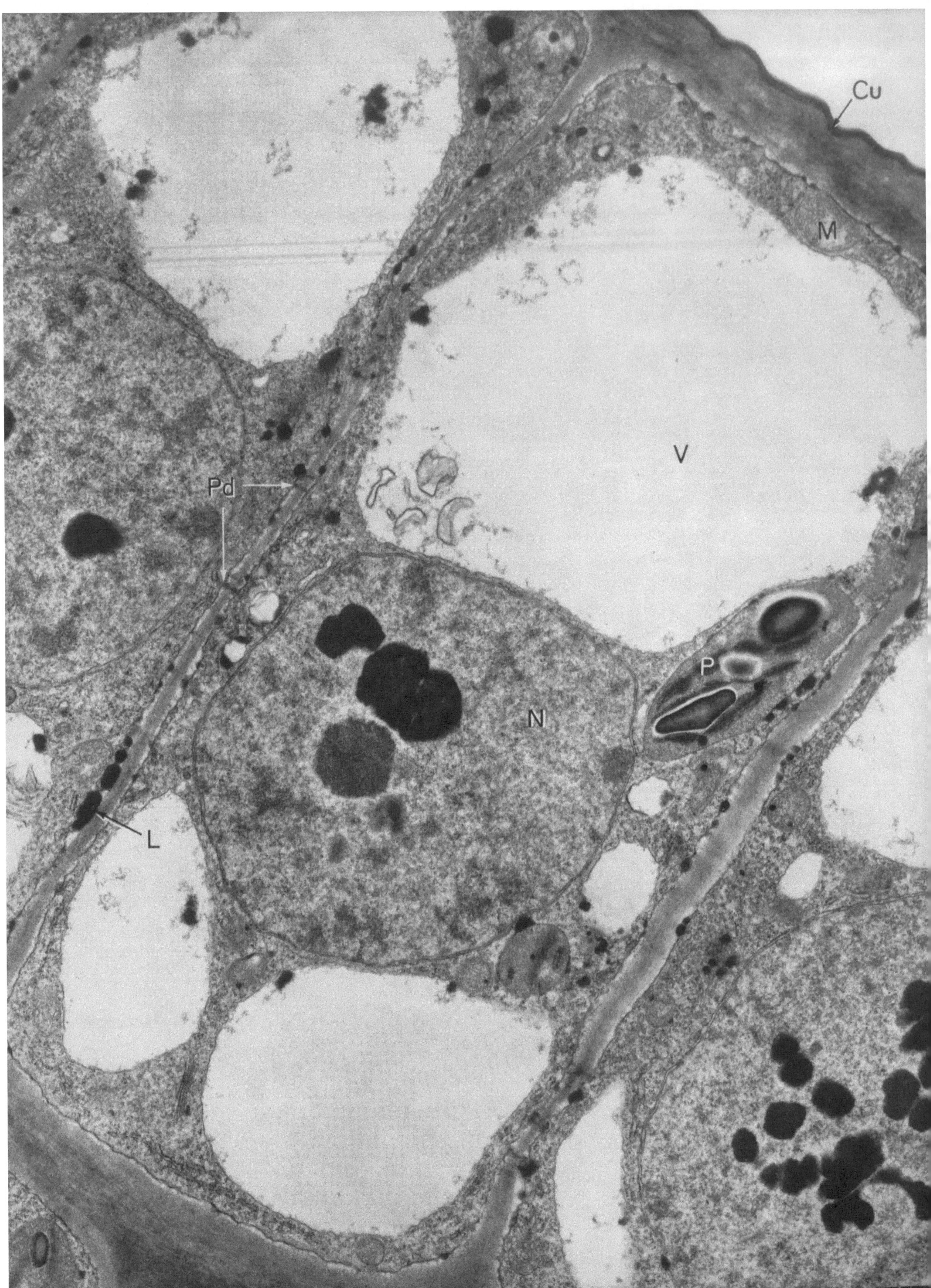

Cu
M
V
Pd
P
N
L

Plate 7.1

Epidermal Cell

The epidermis constitutes a special protective cover for the underlying mesophyll cells of the leaf. As such it appropriately possesses a thick outer wall with a cuticle (Cu) covering its free surface. This, the better to perform its role, is in turn impregnated with or covered by extraordinarily durable waxes (and oils), which essentially waterproof the surface and render it resistant as well to insect pests and the invasion of microbial or viral diseases. The epidermal cells are seemingly specialized for the synthesis and deposition of these important covering materials (Plate 7.2.1).

The example chosen here to illustrate the epidermis is from a young developing leaf of privet. Prior to this stage in their development, the cells of the epidermis had undergone rapid growth along their anticlinal axes. In a later stage of leaf expansion the same cells would have acquired a longer periclinal dimension.

The anticlinal or lateral walls adjacent to other cells of the epidermis may be smooth or wavy, depending on the species. They are usually the thinnest of the walls enclosing these cells and are classified as primary. Though similar in this respect, the wall at the free surface (together with overlying cuticle) is much thicker and is impregnated, possibly with lignin, giving it a greater density. The most superficial layer of this wall, the waxy coat that covers the cuticle, has been removed by preparation procedures (wax solvents).

The cytoplasts are not especially remarkable for their structure or content. Vacuoles (V) are a prominent feature, and mitochondria (M), ER cisternae and plastids (P) are also present. The latter are generally not differentiated for photosynthesis. (The apparent purpose for this is to keep the epidermal cells as clear as possible so that the greater part of the sunlight will move through to the chlorophyll-rich meosphyll.) Starch is commonly stored in the plastids available. The nature of the granules that are evident in the nuclei (N) of these particular cells is not known. The polyhedral outline of these inclusions indicates that they are crystalline and may represent either an intercurrent virus or a storage product. Just outside each protoplast, between it and the lateral walls, there are numerous dense osmiophilic granules (L), which are also scattered in smaller numbers throughout the cytoplasm. These probably represent lipoidal material in transit to the free surface of this tissue. It is worth noting that these droplets accumulate at the anticlinal and apical surfaces of these cells and are virtually absent from the basal surfaces facing the mesophyll tissue of the leaf. Plasmodesmata (Pd) in large numbers extend through the lateral walls to connect adjacent cells.

The epidermal cell layers of the aerial parts of plants are derived from the cells at the surface of the shoot apex, which layer contributes in some instances also to the underlying tissues. Anticlinal divisions and cell expansion allow for the increase in surface necessary for growth of the plant parts. Except for stomata and the pores of hydathodes, there are essentially no interruptions in the continuity of the epidermis. Intercellular spaces are absent except in some petals, and even in such rare instances where these extend to the free surface, the gap is covered by a layer of cuticle. Such epidermal layers are very durable and will persist as long as the organ covered remains alive and active. If a lesion is produced or develops naturally from tissue breakdown, a phellogen or layer proliferating cells appears and proceeds to differentiate into a layer of cork that fills the wound.

From a young leaf of privet, *Ligustrum vulgare* L.

Magnification × 17,000

7.1 Supplemental Reading

Bayley, S. T., Colvin, J. R., Cooper, F. P., Martin-Smith, C. A.: The structure of the primary epidermal cell wall of *Avena* coleoptiles. J. biophys. biochem. Cytol. **3**, 171–182 (1957).

O'Brien, T. P.: Note on an unusual structure in the outer epidermal wall of the *Avena* coleoptile. Protoplasma (Wien) **60**, 136–140 (1965).

Setterfield, G., Bayley, S. T.: Deposition of cell walls in oat coleoptiles. Canad. J. Botany **37**, 861–870 (1959).

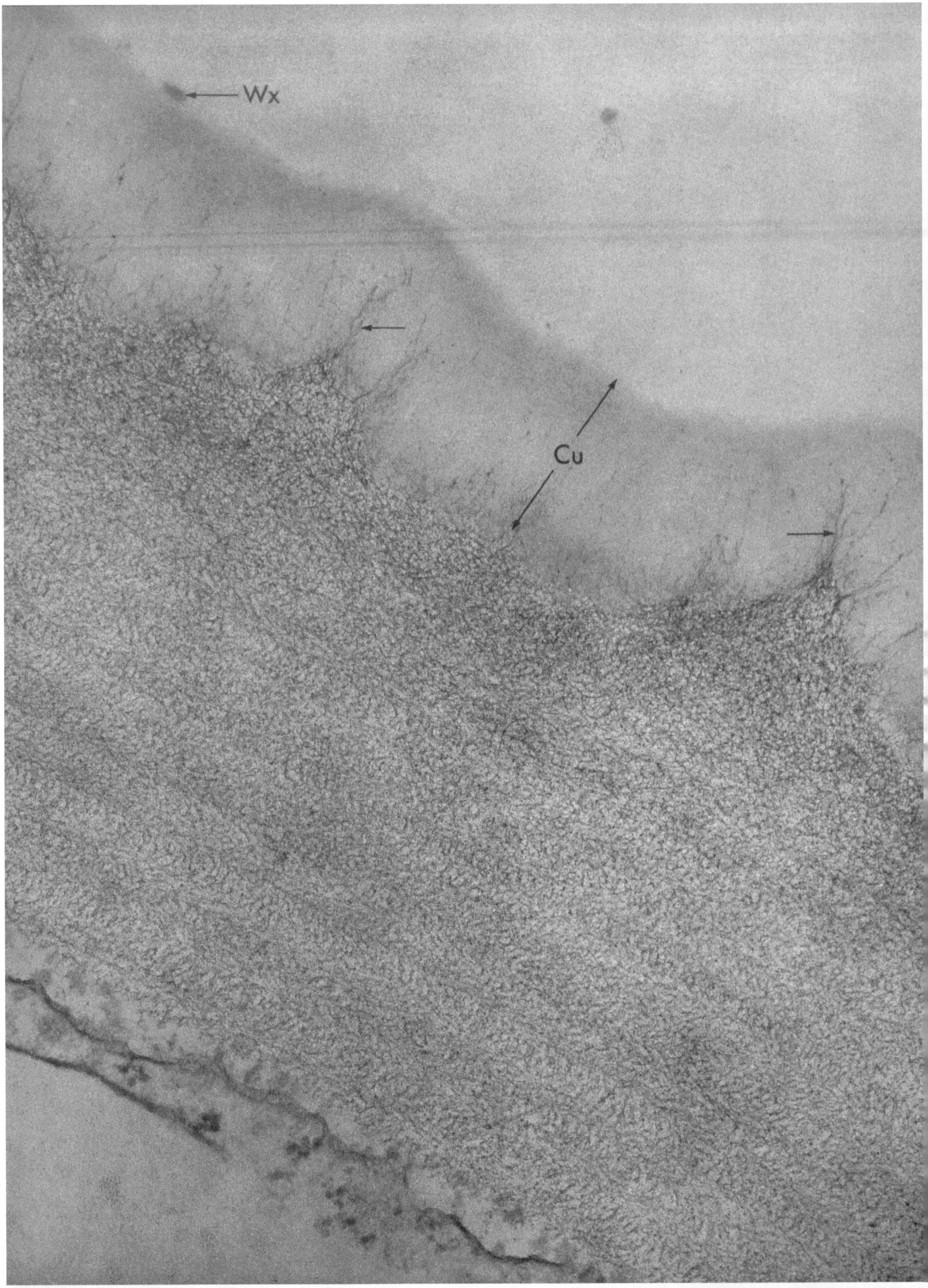
Wx
Cu

Plate 7.2

Epidermal Cuticle of Petiole

As mentioned briefly in the preceding legend, epidermal cells of leaves, stems and fruits are generally covered by a thick cuticular layer. Apparently in the evolution of land plants from aquatic forms, some mechanism was required to prevent excessive loss of water from the cell surfaces. This role seems to have been adopted by the cuticle and its component cutin, for it is universally present over all the exposed surfaces of land plants. Produced by the epidermal cells, it moves through their primary walls and floods out over their exposed surfaces, possibly undergoing some degree of polymerization or curing on exposure to air.

Though cutin is not completely known chemically, the best evidence available indicates that it is a polyester of omega-hydroxymonocarboxylic acids, each containing two straight chains of 16 to 18 carbon atoms. Like the present day polyester plastics, the polymerized cutin is highly resistant to chemical attack and is not known to suffer enzymatic degradation. These properties, among others, doubtless account for much of the resistance shown by the plant surface to the direct attack of pathogens as well as to the wear and tear of time, for it is preserved even in fossil remains.

The cuticle (Cu) depicted here is from the petiole of celery. It is about 300 nm thick as pictured and was probably thicker before exposure to the fat solvents used in tissue embedding. The layer just beneath it appears finely fibrous and doubtless represents the cellulose-rich primary wall. At least it shows a layering that is reminiscent of primary walls (Plate 6.3). Whether the fibrous component stained here represents microfibrils of cellulose or the lignin between them is not known. Either one or the other or both must provide channels or pathways of suitable dimensions and character to facilitate the movement of the monomeric form of cutin to the surface. The slender extensions of this fibrous layer into the superficial cuticle (arrows) are reminiscent of the type of slender channel described by Locke in the insect cuticle and demonstrably involved in wax transport. In plants, the eventual hardening of the cuticle probably terminates the diffusion of cutin. Residual wax on the cuticle surface is indicated at Wx.

The resistance of the cuticle to the passage of water is not complete. Transpiration does apparently take place through these cell surfaces and may be explained by the presence of one or two subterminal hydroxyl groups in the carboxylic acid residues. The cuticle does, however, resist wetting from the outside and so limits the uptake of water by this route.

From a petiole of celery, *Apium graveolens* L. Magnification $\times$ 100,000

7.2 Supplemental Reading

Bystrom, B. G., Glater, R. B., Scott, F. M., Bowler, E. S. C.: Leaf surface of *Beta vulgaris*—electron microscope study. Botan. Gaz. **129**, 133–138 (1968).

Hall, D. M.: The ultrastructure of wax deposits on plant leaf surfaces. II. Cuticular pores and wax formation. J. Ultrastruct. Res. **17**, 34–44 (1967).

Hall, D. M., Jones, R. L.: Physiological significance of surface wax on leaves. Nature (Lond.) **191**, 95–96 (1961).

Juniper, B. E.: The surfaces of plants. Endeavour **18**, 20–25 (1959).

Kolattukudy, P. E.: Biosynthesis of surface lipids. Science **159**, 498–505 (1968).

— Biosynthesis of cuticular lipids. Ann. Rev. Plant Physiol. **21**, 163–192 (1970).

Leyton, L., Juniper, B. E.: Cuticle structure and water relations of pine needles. Nature (Lond.) **198**, 770–771 (1963).

Locke, M.: Permeability of insect cuticle to water and lipids. Science **147**, 295–298 (1965).

Maier, U.: Dendritenartige Strukturen in der Cuticularschicht von *Lilium candidum*. Protoplasma (Wien) **65**, 243–246 (1968).

Schieferstein, R. H., Loomis, W. E.: Development of the cuticular layers in angiosperm leaves. Amer. J. Botany **46**, 625–635 (1959).

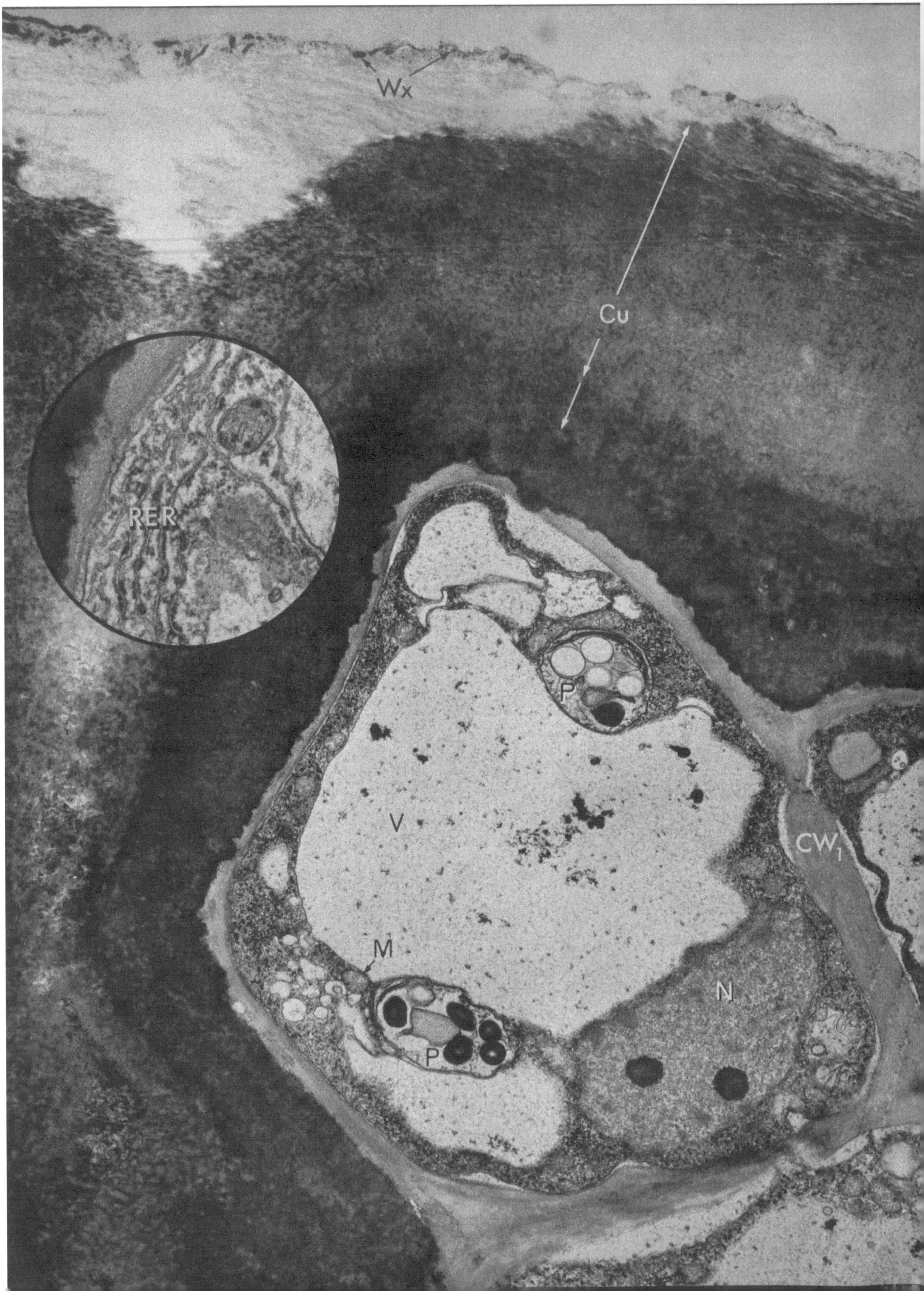
Wx
Cu
RER
P
V
M
P
N
CW_1

Plate 7.2.1

Epidermal Cell and Cuticle of Fruit

In some instances and particularly in fruits, the epidermis is covered by an unusually thick and complex cuticle. Surprising quantities of waxes and cutins exude to the surface and impregnate the available cellulose walls of the cells. Here in the fruit of *Pyrus*, these produce a layer which is one half as thick as the epidermal cells are high. Beneath an outer layer, seemingly fragmented or partly dissolved away and interpreted as a residue of waxes (Wx), there is an extraordinary dense layer (Cu), which doubtless represents some intermediate in the pathway toward the pure waxy coat. Whatever its nature it impregnates part of the primary cellulose wall (CW_1) which lies just subjacent to it. In one or two places this osmiophilic component reaches through to the protoplast surface. Altogether the cuticle shown here is a remarkable surface coat.

The cells of the epidermis underlying this cuticle, even in this mature fruit, appear to retain all the characteristics of normal viable cells. They have a prominent vacuole (V), several plastids (P) containing starch and several small vacuoles, a few mitochondria and profiles of what appears to be unusually well developed rough ER (see insert RER). This is heavily encrusted with ribosomes and for this reason is assumed to be active in the production of enzymes required for the ripening of the fruit. As this latter process continues the cuticle hardens and becomes less and less permeable to all substances including the waxes and oils. The epidermal cells probably go through several phases or sequences in their differentiation, phases involved first in wall production, then cuticle and wax production and finally the production of the free sugars of the ripened fruit.

From the skin of a pear, *Pyrus communis* L. Magnification ×12,000; insert from another epidermal cell of same fruit ×32,000

7.2.1 Supplemental Reading

Chambers, T. C., Possingham, J. V.: Studies of the fine structure of the wax layer of sultana grapes. Aust. J. biol. Sci. **16**, 818–825 (1963).

Eglington, G., Hamilton, R. J.: Leaf epicuticular waxes. Science **156**, 1322–1335 (1967).

Hallam, N. D.: Sectioning and electron microscopy of eucalypt leaf waxes. Aust. J. biol. Sci. **17**, 587–590 (1964).

Sitte, P., Rennier, R.: Untersuchungen an cuticularen Zellwandschichten. Planta (Berl.) **60**, 19–40 (1963).

Skene, D. S.: The fine structure of apple, pear and plum fruit surfaces, their changes during ripening and their response to polishing. Ann. Botany (Lond.) **27**, 581–587 (1963).

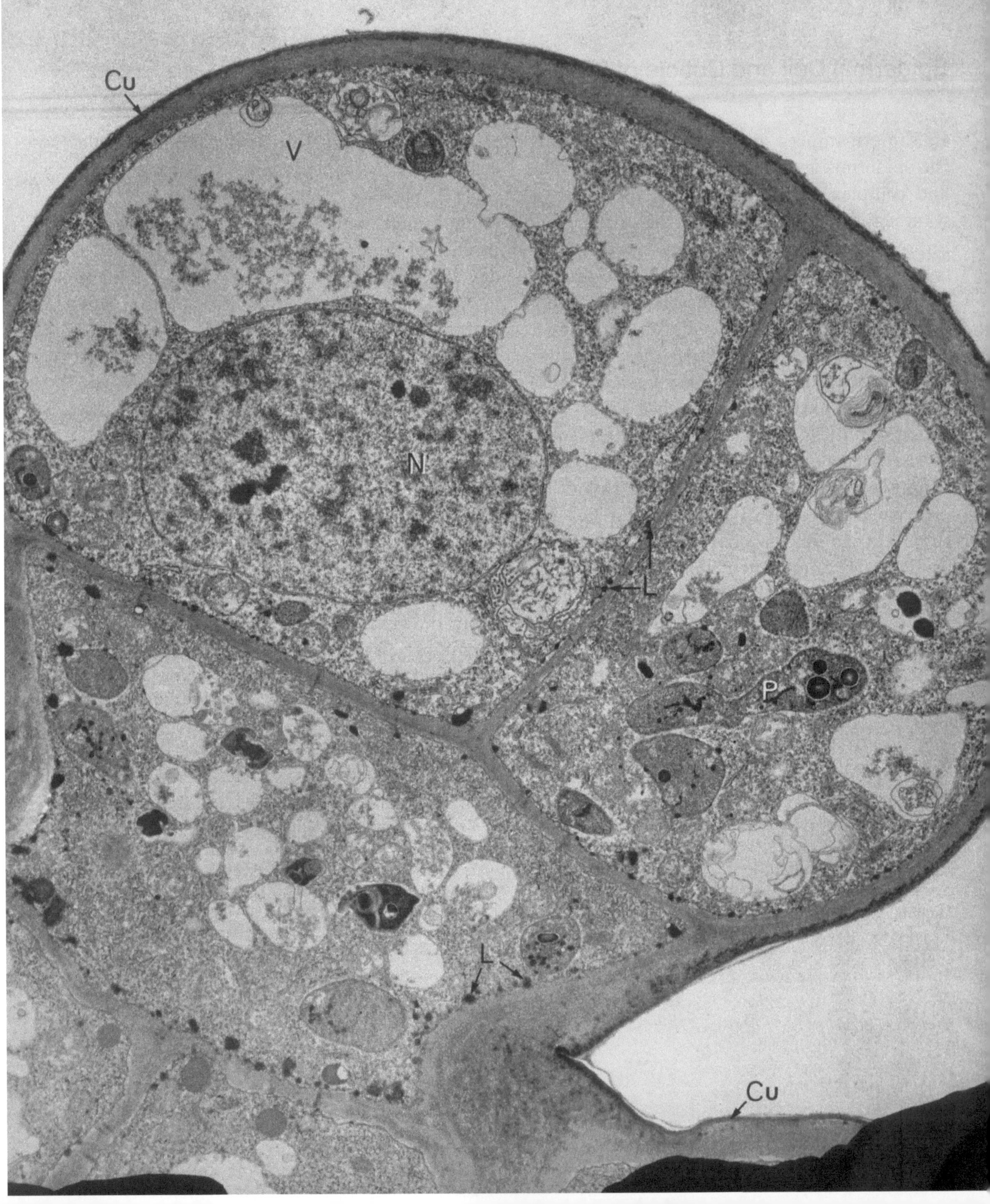
Cu
V
N
L
P
L
Cu

Plate 7.3

Gland Cells

Cells of the plant epidermis show a great diversity in the forms of their free surfaces. Some possess slender extensions (hairs) called trichomes, others are rounded in a manner shown in this micrograph. In this instance the cells function as glands and secrete a variety of substances, mostly terpenes or essential oils. It is these glands then that give to many plant leaves their striking odors and their value as sources of food seasonings.

The gland shown here consists probably of three cells supported on a stalk cell. The whole is surrounded by a cuticle (Cu), which is continuous with the thicker cuticle of the leaf. Primary walls, far less prominent than those over the free surfaces, separate the secretory cells. Vacuoles (V), mitochondria and plastids (P) can be identified within them. The cytoplasm is quite dense, and includes a striking aggregation of ER profiles and particulates (some ribosomes). The nucleus (N), which is large and said to be polyploid, shows an unusual distribution of heterochromatin.

Numerous dense granules occupy a space just beneath the cell wall and are reminiscent of those noted earlier in the cells of the epidermis of privet (Plate 7.1). They presumably represent the lipoidal substances (essential oils?) eventually included in the cuticle.

From the leaf of privet, *Ligustrum vulgare* L. Magnification ×12,000

7.3 Supplemental Reading

Horner, H. T., Jr., Lersten, N. R.: Development, structure and function of secretory trichomes in *Psychotria bacteriophila* (Rubiaceae). Amer. J. Botany **55**, 1089–1099 (1968).

Mollenhauer, H. H.: The fine structure of mucilage secreting cells of *Hibiscus esculentus* pods. Protoplasma (Wien) **63**, 353–362 (1967).

O'Brien, T. P.: Cytoplasmic microtubules in the leaf glands of *Phaseolus vulgaris*. J. Cell Sci. **2**, 557–562 (1967).

Scala, J., Schwab, D., Simmons, E.: The fine structure of the digestive gland of Venus's flytrap. Amer. J. Botany **55**, 649–657 (1968).

Schnepf, E.: Zur Feinstruktur der schleimsezernierenden Drüsenhaare auf der Orchrea von *Rumex* und *Rheum*. Planta (Berl.) **79**, 22–34 (1968).

Shimony, C., Fahn, F. L. S.: Light- and electron-microscopical studies on the structure of salt glands of *Tamarix aphylla* L. J. Linnean Soc. Botany **60**, 283–288 (1968).

Uphof, J. C. Th.: Plant hairs. Berlin: Gebrüder Borntraeger 1962.

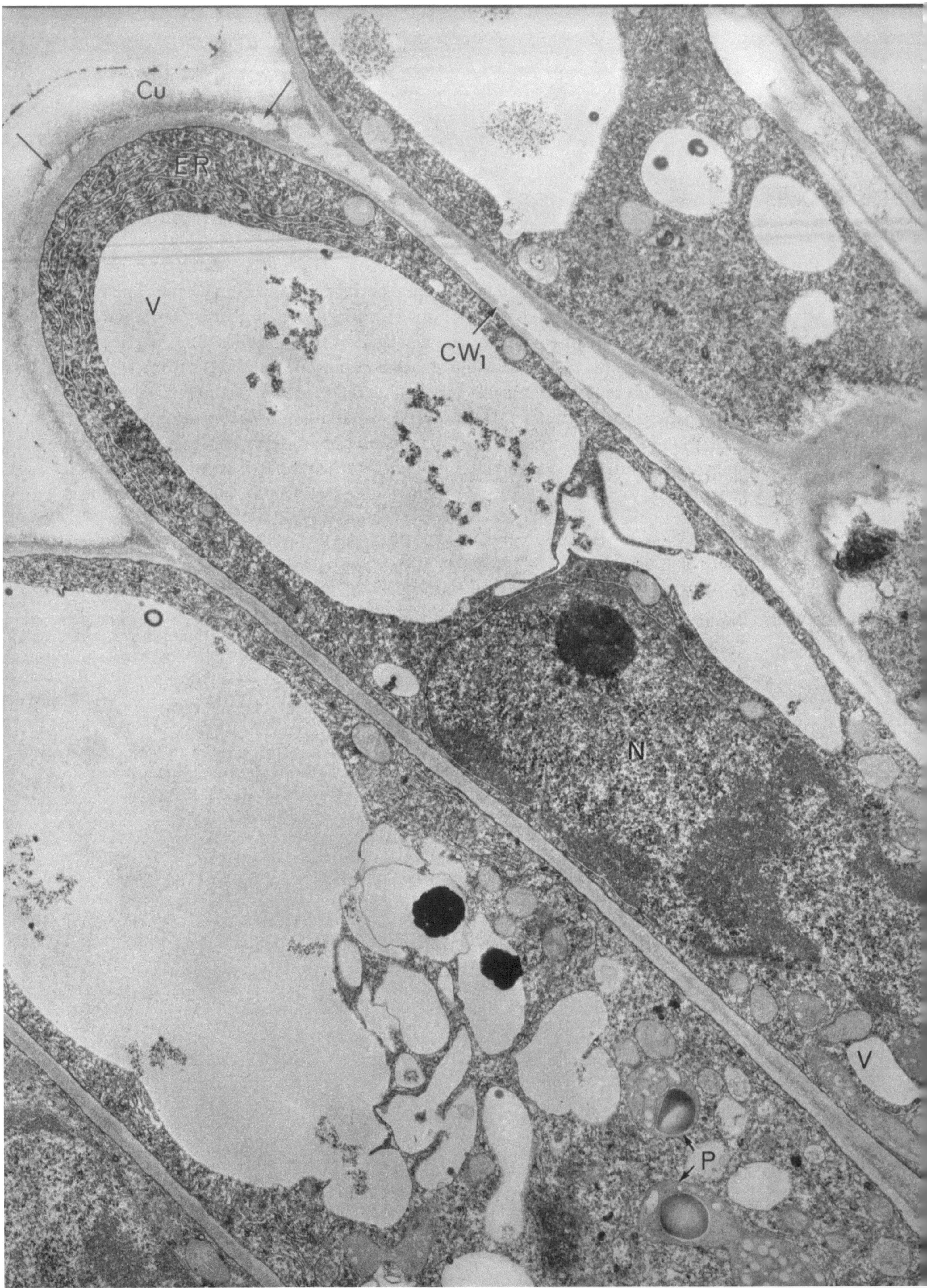

Cu
ER
V
CW1
N
V
P

Plate 7.4

Secretory Cells of a Nectary

Among the several variations in form and function adopted by epidermal cells of the vascular plants, none is more striking than the cells of the nectary. As is well known, these cells produce concentrated sugar solutions (plus specific attractants) which seduce insects to become involved in pollination. These cells are generally located somewhere basal to the gynoecium of the flower, and the product of their secretion accumulates at this point.

The nectary cells shown here are from the Crown-of-Thorns. They are tall columnar units and more highly polarized than most plant cells. Each is separated from its neighbor by a thin primary wall (CW_1) and each rounds out at its free surface into a small crown. There is a thick cuticle (Cu) over this surface, but it seems not to represent a barrier to the diffusion of secretory products from these cells. A prominent vacuole (V) occupies the apical half of the cell and seems to force the nucleus (N) into a basal position. The layer of cytoplasm surrounding the vacuole is remarkable in being quite thick, in showing a level of density that equals that of the nucleus (N) and in possessing as compact an array of thin ER cisternae (ER) as one encounters anywhere in plant cells. There is no ready explanation for this or any other of these features, but presumably they are designed to support in some way the active production of nectar. The concentration of ER elements just under the apical surface would seem to be especially significant in this regard.

In other respects these nectary cells are typical of epidermal cells. They show small plastids (P), containing starch and small vacuoles. Probably these stores of carbohydrate are mobilized in periods of secretion. Other less presumptive evidence relates the quality of the secretion (sugar conc.) to the proximity and character of the vascular tissue.

By what device the product of secretion moves through the cuticle is not evident in this micrograph. There are light microscope studies which depict tiny canals going from the cell surface to the free surface of the cuticle. None was observed in these electron micrographs, but it may be that none was luckily caught in the section. The only unusual thing is the obvious space between cells at the top and the extension of the intercellular spaces to the region just beneath the cutucle (arrows). Conceivably this space channels to the free surface the sugar diffusing from the cells and from underlying vascular elements. From a nectary of Crown-of-Thorns, *Euphorbia Milii* Ch. des Moulins. Magnification × 16,000

7.4 Supplemental Reading

Nieuwenhuis von Uexkull-Güldenband, M.: Sekretionskanäle in den Cuticularschichten der extrafloralen Nektarien. Rec. Trav. Botan. Neerlandais **11**, 291–311 (1914).

Schnepf, E.: Zur Cytologie und Physiologie pflanzlicher Drüsen. 5. Teil: Elektronenmikroskopische Untersuchungen an Cyathialnektarien von *Euphorbia pulcherrima* in verschiedenen Funktionszuständen. Protoplasma (Wien) **58**, 193–219 (1964).

Schnepf, E.: Sekretion und Exkretion bei Pflanzen. Protoplasmatologia Bd. VIII, Tl. 8. Wien: Springer 1968.

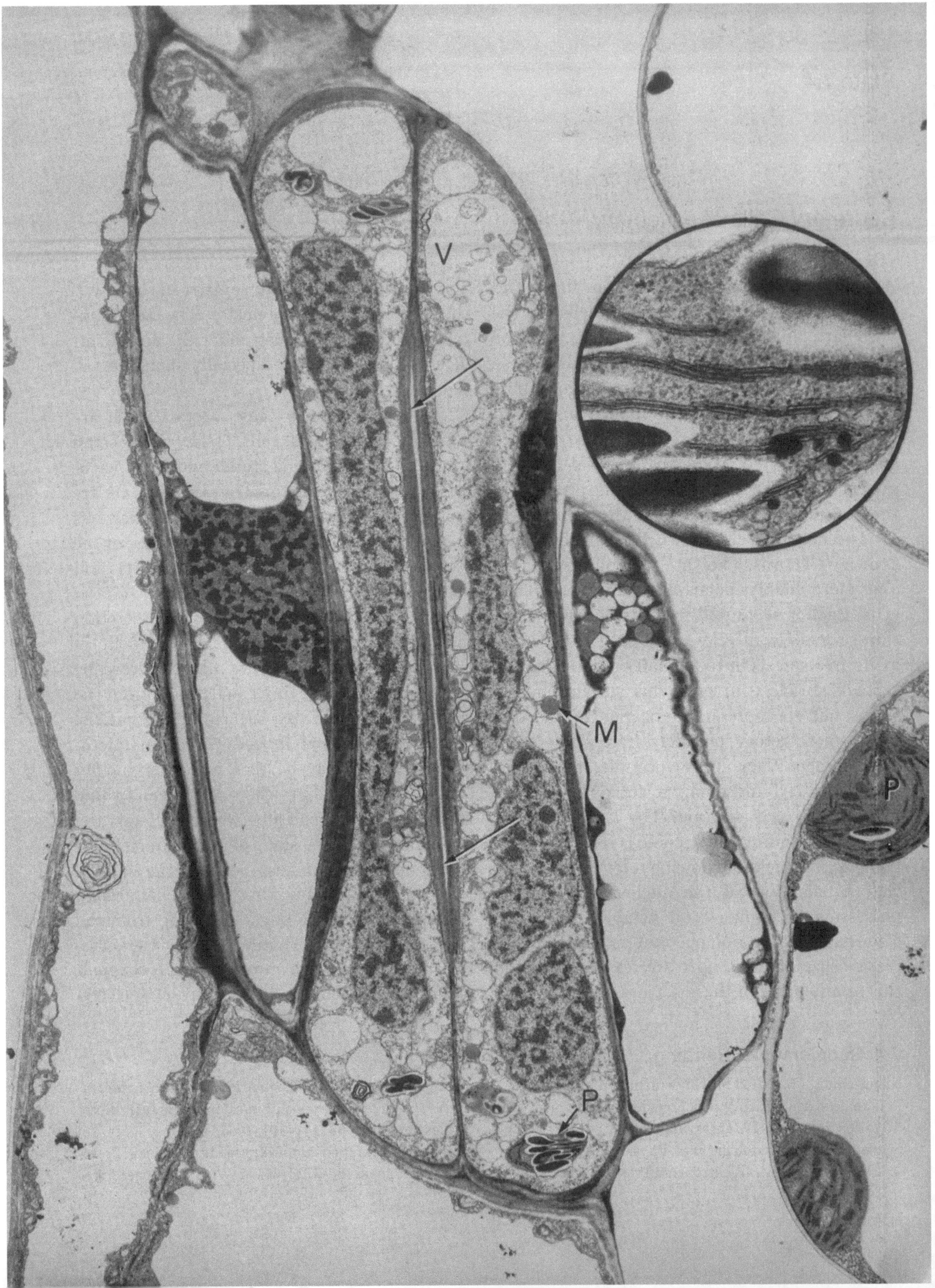
V
M
P
P

Plate 7.5

Leaf Stoma

The small groups of four cells which make up the stomatal apparatus are, functionally speaking, among the most unusual found anywhere in plants. They occur most commonly in leaves as part of the abaxial epidermis but are encountered in smaller numbers in the surface layers of almost all other parts of most plants. Actually the density of stomata in leaves may reach several hundred per square millimeter of epidermal surface.

This micrograph shows a section cut parallel to the surface of a leaf of *Phleum pratense*. At center it includes two long guard cells that limit a slit-like opening, the stoma (arrows). Lateral to the guard cells the section cuts through subsidiary cells and thence into epidermal units on the left and mesophyll cells on the right. The same kinds of cells and their relationships are shown in another view in Fig. 7.5.1.

As is well known, stomata open and close in response to changes in the composition (particularly H_2O and CO_2) of the surrounding air and to light of wavelengths that are most effective in photosynthesis. Thus they function to control the rate of gas exchange between the plant interior and its external environment. Changes in the aperture size result from shifts in turgor pressure within the cells, which in turn reflect differences in osmotic pressure between the guard cells and their adjacent subsidiary cells. An increase in turgor resulting from water uptake generally opens the stomata; a decrease, such as that accompanying periods of desiccation generally closes them. The minor changes in shape which guard cells make in response to shifts in turgor pressure are influenced by the relatively enormous variations in wall thickness over the cell's own surface. Though these variations in wall design range widely from species to species of plant, there is constancy in each species. Guard cells are uniform in possessing a thickened ledge along the closing edge to insure a tight seal when the aperture is closed. A space, a substomatal chamber, is present below the aperture (see Plates 7.5.1 and 7.5.2).

The guard cells of *Phleum pratense* are fairly typical. They are living and nucleated. The distribution of heterochromatin in small dense masses is quite distinctive. The cytoplasm is marked by many small vacuoles (V), all part of the tonoplast system. This probably plays a role in the development of turgor pressure and toward that end may contain relatively high concentrations of glucose. Mitochondria (M) are dense and quite numerous for a plant cell that is not active in synthesis. Their presence suggests, then, that control of osmotic pressure involves the metabolism of energy-rich compounds. The plastids (P) are small compared with those in the adjacent mesophyll cells, but they are loaded with starch, here seen as dark granules, representing possibly an available source of metabolizable carbohydrate.

Characteristic variations in wall thickness are evident in this image, but they are more completely understood for the whole cell by reference to Plate 7.5.1. From a marked thickness of several micrometers along the aperture, the wall thins out to a fraction of a micrometer between the guard cells at their ends. This thinning is thought to facilitate rapid equilibration of turgor between the two cells, a process further abetted by the presence of plasmodesmata. Similar thin areas of primary wall face the subsidiary cells, which have very large vacuoles and presumably serve as a ready source of H_2O. This relationship between the guard and subsidiary cells is better illustrated in the cross section shown in Plate 7.5.1.

The plastids (P) that are present in guard cells would seem to function as photoreceptor units and as carbohydrate producers. That they are equipped with a lamellar organization (stacks and grana) sufficient for photosynthesis is indicated in the enlargement of one, shown in the insert at the upper right. The number of lamellae per stack, while not great, is typical of small grana in other cells where photosynthesis occurs. Presumably the starch precursors are synthesized locally and not imported from other cells of the leaf.

From a leaf of timothy, *Phleum pratense* L.
Magnification $\times$8,000
Insert $\times$65,000

7.5 Supplemental Reading

Brown, W. V., Johnson, S. C.: The fine structure of the grass guard cell. Amer. J. Botany **49**, 110–115 (1962).

Meidner, H., Mansfield, T. A.: Physiology of stomata. New York: McGraw Hill 1968.

Zelitch, I.: Stomatal control. Ann. Rev. Plant Physiol. **20**, 329–350 (1969).

— Control of leaf stomata—their role in transpiration and photosynthesis. Amer. Scientist **55**, 472–486 (1967).

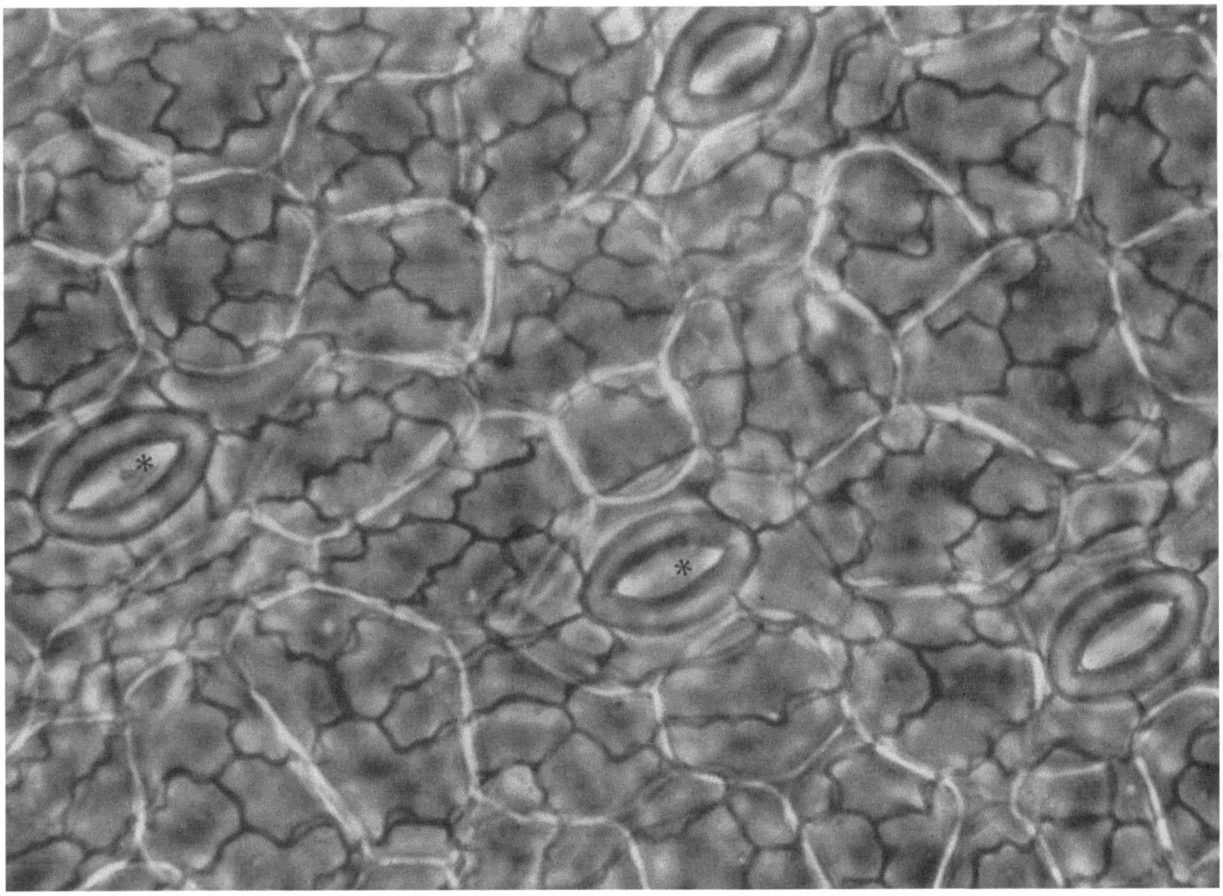

Text Fig. 7.5a

Leaf Epidermis, Face View

This phase light micrograph depicts the distribution and form of both stomata and their guard cells as they occur in the leaf epidermis of duck weed, the source also used for the guard cells in Fig. 7.5.2.

Two of the stomata in the image are marked by asterisks. They are limited by sausage-shaped guard cells, which have complex thickenings along their inner margins (see Fig. 7.5.2). The dense, wavy lines in the areas of the picture surrounding the stomata represent the lateral walls of cells making up the epidermis, whereas the paler outlines that intermingle with these mark the limits of mesophyll cells beneath the epidermis.

From the frond of duck weed, *Spirodela oligorrhiza* Kurz.

Magnification $\times 720$

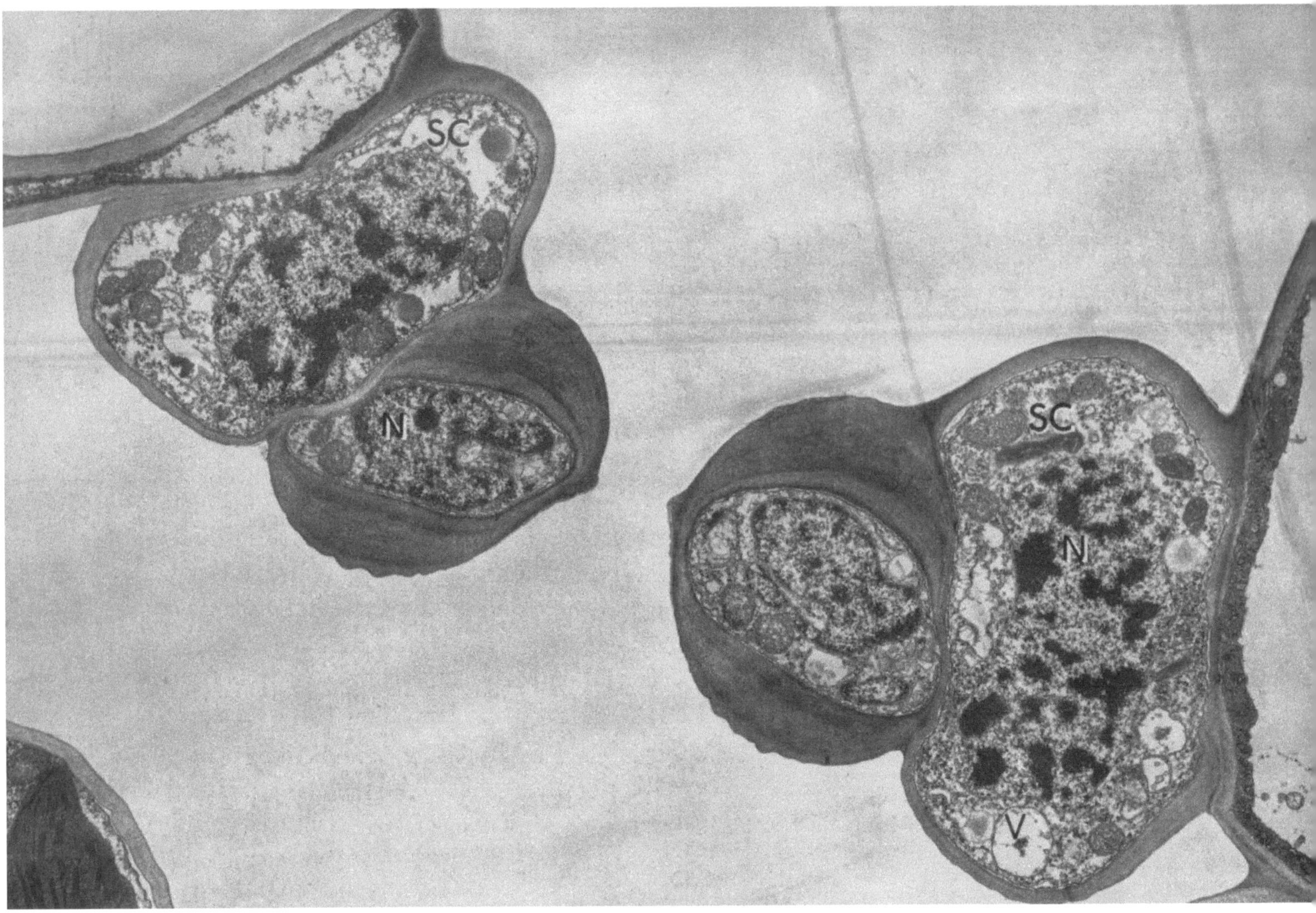
SC
N
SC
N
V

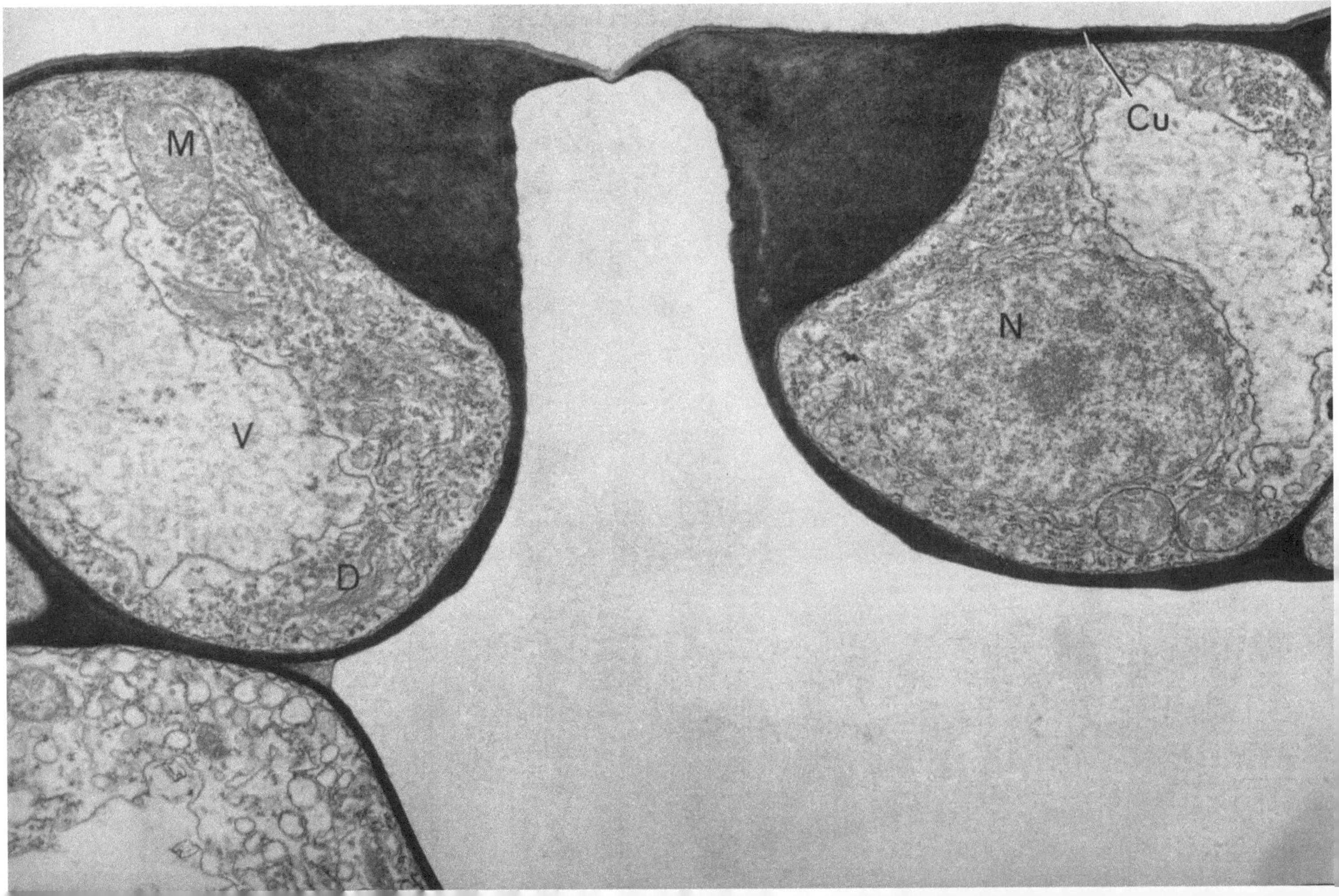
M
V
D
Cu
N

Fig. 7.5.1 (upper figure on page 128)

Guard Cells of Stomata

The four cells which make up the stomatal complex in *Phleum pratense* are depicted here in cross-section. The central two, separated by a gap, are the guard cells; the pair lateral to them the subsidiaries (SC); and beyond these are parts of adjacent epidermal cells. The one at the upper left has collapsed in preparation.

From this section, cut medially within the complex, one might judge that the guard cells are smaller than the subsidiaries. Actually this is not the case. The guard cells are simply longer and narrower. They are also different from the subsidiaries in having both thicker and thinner zones in their surrounding wall. These differences influence the change in shape of the guard cells in response to shifts in turgor pressure and secondarily affect the opening and closing of the aperture. It is obvious that the subsidiary cells are not so constructed. Furthermore, reference to Fig. 7.5.2 makes it evident that the thickenings in the wall of the guard cells need not be *Phleum*-like to achieve an effective control. In actual fact many varieties of guard cells, especially in terms of wall sculpturing, have evolved to take care of this same function.

The morphogenesis of these highly specialized forms has excited a lot of interest, for it is clear that some intracellular plan of organization that is under genetic control, must determine where the thicker and thinner parts of the wall will be deposited. In the guard cells this pattern occurs in mirror image. Within recent years the interest in this phenomenal differentiation has been stimulated by the discovery that during wall thickening cytoplasmic microtubules are most abundant in the cytoplast cortex underlying the zones of most active cellulose deposition. These seem in turn to be controlled in their distribution and orientation by sites in the cytoplasmic ground substance from which they take their origin. From a leaf of timothy, *Phleum pratense* L. Magnification $\times$ 8,500

Fig. 7.5.2 (lower figure on page 128)

(lower figure on page 128)

Guard Cells of Stomata

This micrograph illustrates further the variation in form that guard cells may adopt. The image derived from this cross section of a stoma of *Spirodela* includes a profile view of the ledge, i.e., the thickenings along the facing surfaces of the guard cells, which meet when the stoma closes (Plate 7.5). The density of the secondary thickening, and in fact the entire wall, suggests a high degree of lignification. A thin layer of material interpreted as cuticle (Cu) covers the exposed surface.

Just how these particular guard cells work is difficult to suggest. The lips or ledges of the closing margins could be separated by several different distortions of the cells, including one that would move them like swinging doors.

Besides differing in wall form from the guard cells in *Phleum* (see Fig. 7.5.1), these cells in *Spirodela* show a somewhat different cytoplasmic organization. There is more rough ER, and the tonoplasts (V) are not so extensively divided.

From a frond of duck weed, *Spirodela oligorrhiza* Kurz.
Magnification ×15,000

7.5.1 and 7.5.2 Supplemental Reading

Heath, O. V. S., Mansfield, T. A.: The movements of stomata. In: The physiology of plant growth and development, p. 301–332. (Wilkins, M. B., ed.) New York: McGraw Hill Book Co. 1969.

Ketallapper, H. J.: Stomatal physiology. Ann. Rev. Plant Physiol. **14**, 249–270 (1963).

Pickett-Heaps, J. D., Northcote, D. H.: Cell division in the formation of the stomatal complex of the young leaves of wheat. J. Cell Sci. **1**, 121–128 (1966).

Stebbins, G. L., Jain, S. K.: Developmental studies of cell differentiation in the epidermis of monocotyledons. I. *Allium, Rhoeo*, and *Commelina*. Develop. Biol. **2**, 409–426 (1960).

— Shah, S. S.: Developmental studies of cell differentiation in the epidermis of monocotyledons. II. Cytological features of stomatal development in the Gramineae. Develop. Biol. **2**, 477–500 (1960).

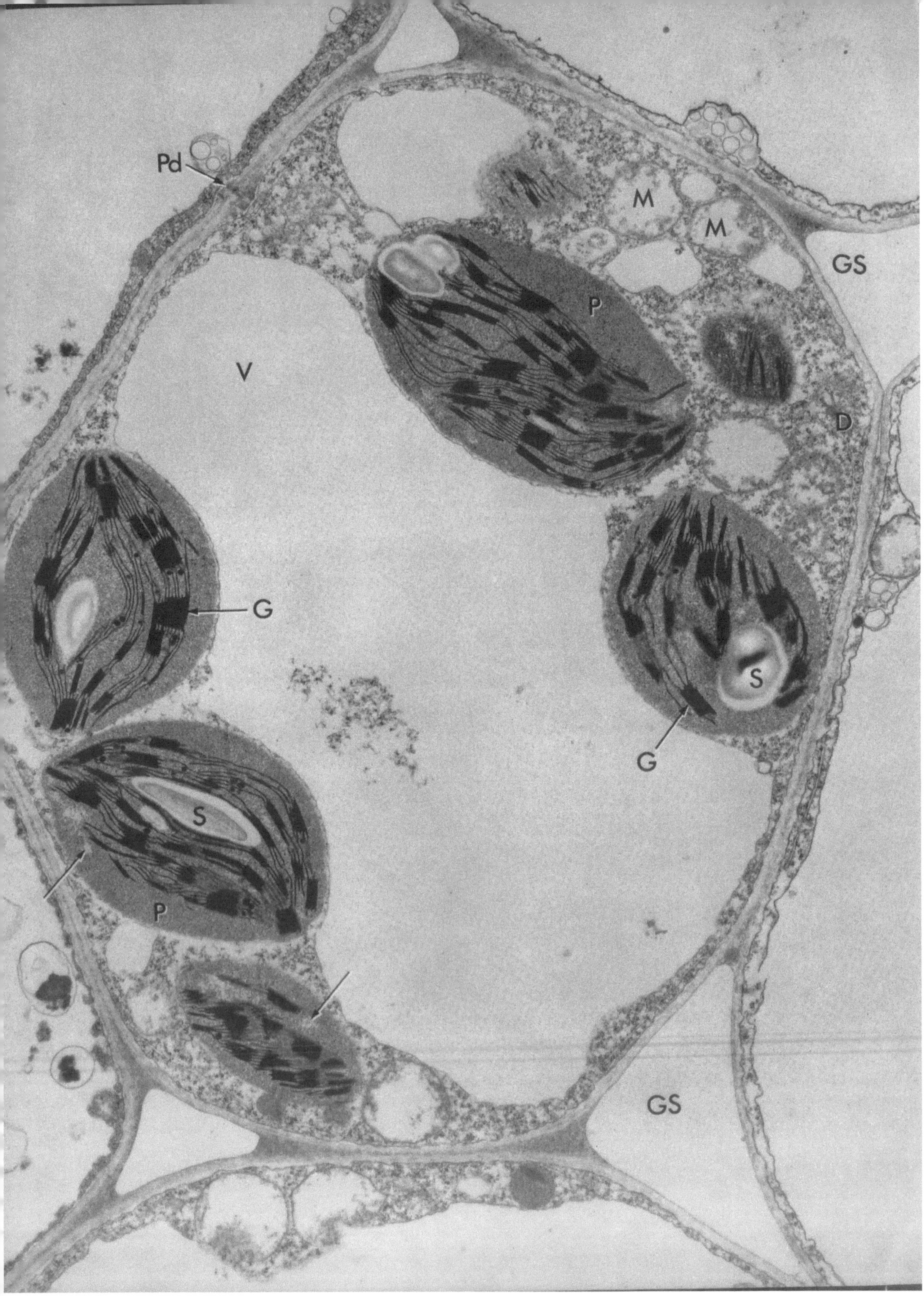
Pd
M
M
GS
P
V
D
G
S
G
S
P
GS

Plate 8.1

Mesophyll Cell

Mesophyll tissues of leaves frequently occur in two arrangements, which are referred to as palisade and spongy parenchyma. Where it exists, the palisade type is made up of columnar or prismatic cells tightly packed with their long axes parallel, an arrangement thought to contribute to the efficiency of the photosynthetic process. The spongy type, on the other hand, is composed of loosely arranged and loosely attached cells with large intercellular spaces. This type of structure is well adapted to the gaseous exchanges that are essential to photosynthesis and respiration in this tissue.

The mesophyll tissue shown in this micrograph is from *Phleum* and is typical for members of the grass families, in which a clear distinction between the palisade and spongy arrangements is not evident. It would seem that in these plants the mesophyll cells are as well adapted to photosynthesis as to gas exchange.

The large size of mesophyll cells is due in part to their relatively large vacuole (V), and in part also to the large number of big lens-shaped chloroplasts (P) they contain. These starch-rich (S) organelles are confined to the layer of cytoplasm between wall and vacuole. As a rule their shortest axis is oriented normal to the plane of the adjacent cell surface. Whether they are held in this position by the thinness of the cytoplasmic layer in which they reside and move (a space ordinarily not large enough to accommodate the longer axes), or whether some other mechanism is operative in this orientation is not known. It is interesting and possibly significant for their role in photosynthesis that the planes occupied by the lamellae and grana of the chloroplasts are parallel to the long axis of the cell (here vertical to the page) and thus to the lateral surfaces of the cells.

Vacuoles, smaller than the major central unit (V), are present in small number in the peripheral cytoplasm. Nearby mitochondria (M) have suffered some damage from preparation procedures. Large lamellar cisternae of the endoplasmic reticulum hug the cell surface. Dictyosomes (D) are rare in these cells, so it seems that little if any secretion of large polysaccharides or mucoproteins is likely to occur.

The spongy character of this mesophyll tissue is exemplified by the large intercellular spaces. These gas-filled cavities, which in some leaves account for as much as 70% of the leaf volume, are evident here at the cell corners (GS). They constitute a labyrinth that surrounds the cells and is continuous with the spaces under the stomata. The cell walls bordering these spaces are in each instance quite thin (ca. 0.1 μm) and lack any evident lignin or encrusting substances. These facts plus the known hydration of these walls suggests that they represent little or no barrier to the exchange of O_2 and CO_2 between cell and air spaces. The nature of these walls also permits substantial water loss into the air spaces, the final event in the process of transpiration. The water and salts which make up the transpiration stream come from the xylem elements of the vascular traces. Usually these are only a few cells away from any mesophyll element.

The water must pass from cell to cell through areas of contact or along the walls to points where these walls face air spaces.

The products of photosynthesis, which are stored temporarily in the chloroplasts of mesophyll cells, are eventually passed as small molecules to the sieve elements of the vascular traces for transport to other portions of the plant. It is assumed on the basis of structural relationships of mesophyll cells to other living cells in the leaf, including phloem cells, that the plasmodesmata may provide a continuous protoplasmic phase along which carbohydrates can move preferentially and without any restraints except those imposed by the small cross-sectional area of such connections (see Pd in this micrograph).

The details of chloroplast structure are discussed relative to a succeeding plate (8.2). Here attention is directed only to the grana (G), which stand out prominently, and to the less prominent and poorly defined patches of low density in the chloroplast matrix, which probably contain the DNA now known to be present in all chloroplasts. These areas are called nucleoids (arrows).

From the leaf of timothy, *Phleum pratense* L.
Magnification × 17,000

8.1 Supplemental Reading

Calvin, M.: The path of carbon in photosynthesis. Science **135**, 879–889 (1962).

Gibor, A., Granick, S.: Plastids and mitochondria: inheritable systems. Science **145**, 890–897 (1964).

Green, P. B.: Cinematic observations on the growth and division of chloroplasts in *Nitella*. Amer. J. Botany **51**, 334–342 (1964).

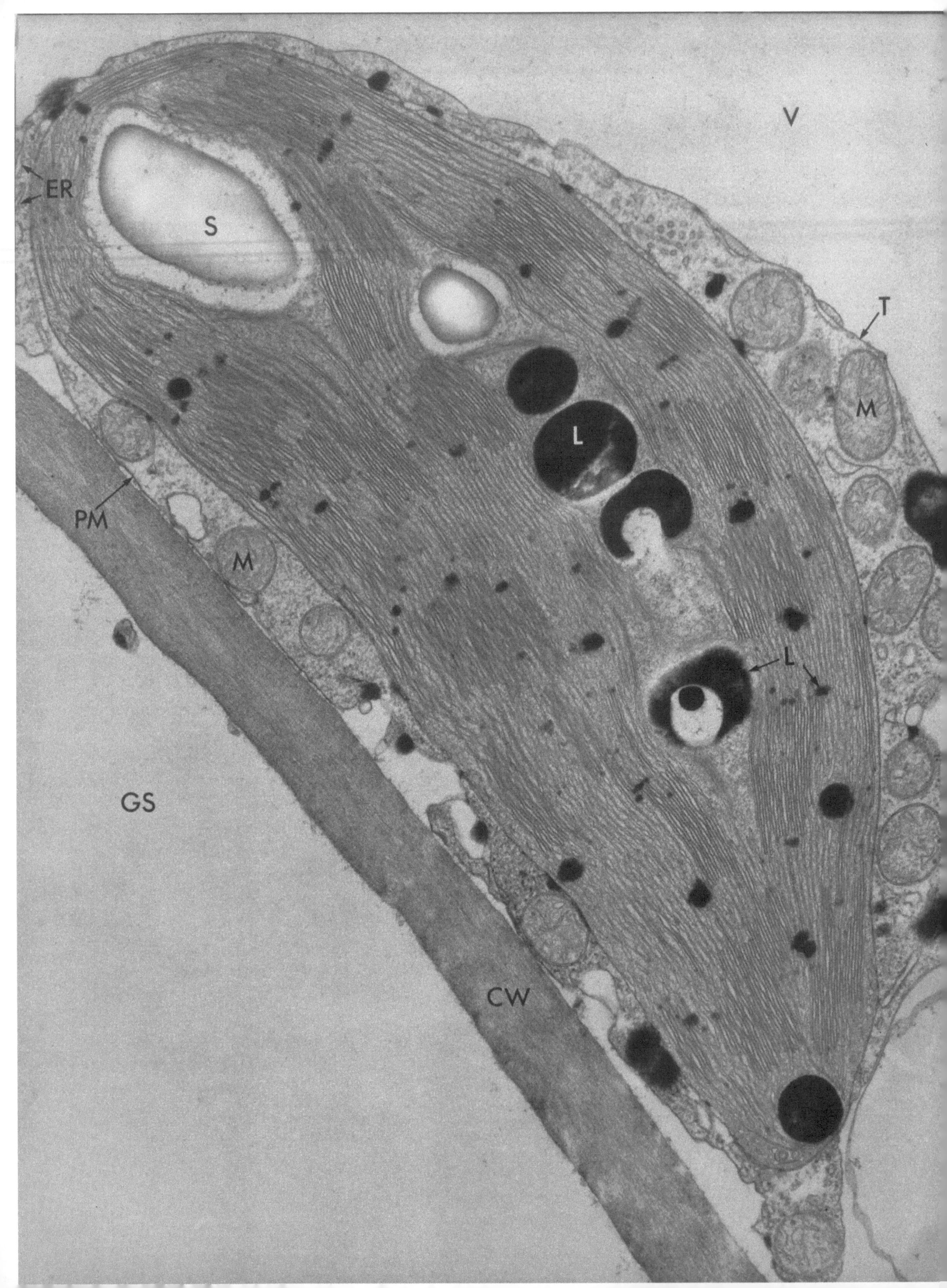
V
ER
S
T
M
L
PM
M
L
GS
CW

Plate 8.2

Chloroplast

The photosynthetic apparatus of plants, the chloroplast, has been intensively studied because of its extraordinary importance. One need only be reminded that all of the foodstuffs, the source of energy for the life and activities of the biological world, not to mention the fossil fuels and the relatively high O_2 content of the earth's atmosphere, are the products of photosynthetic activity occurring in those chlorophyll-bearing structures. In the overall reaction, the energy of light is used to remove electrons from water for the generation of a strong reductant, which reduces CO_2 to carbohydrates. Oxygen is evolved from water as a by-product. These reactions are the reverse of respiration, in which useful energy is derived from the oxidation of carbohydrates with the consequent production of carbon dioxide and water.

As implied above, chlorophyll pigments essential to photosynthesis are most frequently localized within chloroplasts or plastids. The only exceptions are certain photosynthetic bacteria and the blue-green algae. In the higher plants the functioning plastids (chloroplasts) have a relatively uniform morphology and size. In most instances they are shaped like biconvex lenses with diameters of 4–6 μm. The chloroplasts in Plate 8.1, shown in vertical section, are a reasonable example of chloroplasts in general.

As depicted here in a mesophyll cell of *Crassula* the single plastid is surrounded by a layer of cytoplasm containing mitochondria (M), elements of the endoplasmic reticulum (ER) and numerous ribosomes. On one side this cytoplasm is limited by the plasma membrane (PM) of the cell, and on the other (top right) by the tonoplast (T), the limiting membrane of the vacuole (V). Thus the relatively huge chloroplast is sandwiched between cell wall (CW) and vacuole (V) and is held in this position with its longest axis parallel to the cell surface.

Plastids, like mitochondria, are bounded by a double membrane. The outer of these is smooth and forms a thin barrier between the chloroplast's contents and the surrounding cytoplasm. The inner membrane, while similar in thickness to the outer, differs in being connected at a few points with the extensive system of membranous lamellae that are characteristic of the mature chloroplast. In fact, in the early development of the chloroplast lamellae, the inner membrane seems to fold inward and thereby provide the primers, as it were, for the subsequent growth of this system.

Starting from a relatively small organelle or proplastid in the dark-grown cell, the chloroplast grows as this internal system of membranes expands. The basic structural unit of this system appears to be a very thin flattened vesicle called a thylakoid. Individual vesicles then stack up in various ways so that particular areas on their surfaces bind to like areas on adjacent thylakoids. The resulting stacks, which may contain anywhere from a few to 20 or 30 units, were called grana in light microscope and early biochemical studies. These subunits of chloroplast structure are the sites of the photochemical reactions of photosynthesis. They contain the chlorophyll in association with lipoprotein. The individual lamellae in the grana are coextensive in at least one direction with intergranal lamellae, which therefore can be seen to run from granum to granum.

All of these membranous sacs are surrounded by or embedded in the continuous phase of the chloroplast, its matrix or stroma. The enzyme ribulose 1.5 diphosphate carboxylase is responsible for primary CO_2 fixation and is present in large amounts in the leaf. It appears that this enzyme and other enzymes involved in photosynthesis probably constitute a considerable portion of the stroma material. In addition, it is in this matrix that one may identify in growing chloroplasts both ribosomes (Plates 8.2.1 and 8.2.2) and a nucleoid containing DNA. More prominent inclusions of the chloroplast shown in this image include osmiophilic lipid globules (L), both large and small, and starch grains (S), which take up practically no osmium. These latter inclusions fluctuate in number and prominence with periods of photosynthesis.

Evidence is now accumulating to support the conclusion that chloroplasts are self-perpetuating cell organelles, having a supply of genetic information. This in turn determines the characteristics of the many proteins and other compounds which make up these extraordinary cell inclusions. The interaction of environmental factors, such as light, with information programmed into the young chloroplast triggers synthetic processes that are essential for photosynthesis.

From a leaf of *Crassula argentea* Thunb.
Magnification $\times$ 32,000

8.2 Supplemental Reading

Criddle, R. S.: Structural proteins of chloroplasts and mitochondria. Ann. Rev. Plant Physiol. **20**, 239–252 (1964).

Gunning, B. E. S.: The greening process in plastids. I. The structure of the prolamellar body. Protoplasma (Wien) **60**, 111–130 (1965).

Hoffman, L. R.: Observations on the fine structure of *Oedogonium*. III. Microtubular elements in the chloroplasts of *Oe. cardiacum*. J. Phycol. **3**, 212–221 (1967).

Kirk, J. T. P., Tilney-Bassett, R. A. E.: The plastids. Their chemistry, structure, growth and inheritance. San Francisco: W. H. Freeman, Co. 1967.

Lichtenthaler, H. K.: Plastoglobuli and the fine structure of plastids. Endeavour **27**, 144–149 (1968).

Pickett-Heaps, J. D.: Microtubule-like structures in the growing plastids of chloroplasts of two algae. Planta (Berl.) **81**, 193–200 (1968).

Woodcock, C. L. F., Fernandez-Moran, H.: Electron microscopy of DNA conformations in spinach chloroplasts. J. molec. Biol. **31**, 627–631 (1968).

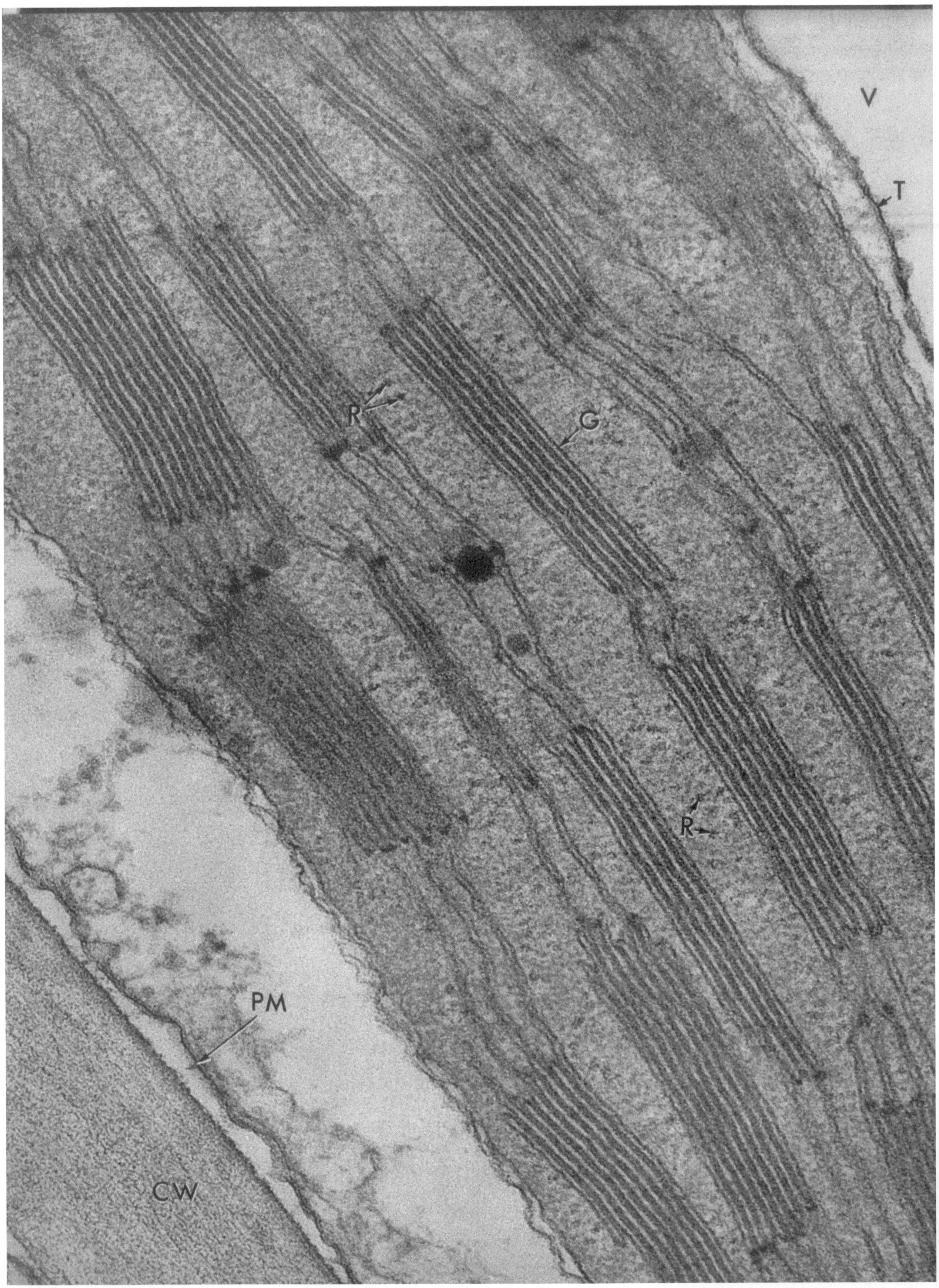
V
T
R
G
R
PM
CW

Plate 8.2.1

Fine Structure of a Chloroplast

As noted in the legend for the preceding plate, the photosynthetic pigments of green algae and the vascular plants are in some way built into the thin flattened sacs or lamellar vesicles known as thylakoids. The membranes of the thylakoids serve as essential barriers across which, during photosynthesis, electrical potentials and pH differences may arise resulting in ion pumping from one side to the other and the formation of adenosine triphosphate from from adenosine diphosphate. The thylakoids measure not thicker than 15 nm. They always lie with their long dimension parallel to that of the chloroplast. Special marginal regions of the thylakoids associate face to face with similar areas of other thylakoids to form tight stacks, the so-called grana (G) of the chloroplast. These may be up to 600 nm in diameter. Because of this origin and construction, the grana are connected through the stroma of the chloroplast by thin membrane-bounded cisternae, called the stroma lamellae, intergranal lamellae (text Figs. 8.2.1a and 8.2.1b) or "frets". Even a casual study of a vertical section through a granum shows the membranes and spaces descriptive of this structure. For example, if we start with the dense line at G, which represents the edge-on view of a membrane (the characteristic trilaminar structure of the "unit membrane" is not resolved), we see immediately adjacent to it a thin (4 nm) space of low density, then two membranes closely apposed and separated only by a thin (1–1.5 nm) line of low density, and thereafter another intrathylakoid space and so on repetitiously across the depth of the granum. This association, into tightly packed arrays, of the membranes bearing the photosynthetic pigments is thought to contribute some efficiency to the photochemical mechanisms by which light energy is captured and transformed into chemical energy for use by the plant. The precise character of this structural-functional relationship is not as yet understood.

An additional feature of the granum lamellae is worth noting. At their free edges there is usually a small knot of high density which is thought to be a small concentration of lipids.

Text Fig. 8.2.1a is an image of thylakoid and granum structure that may appear after the use of glutaraldehyde followed by osmium and is strikingly different from that in 8.2.1 following the same preparation procedure. No explanation for this phenomenal appearance (and difference) has been provided, and it is difficult to refer it to any functional state of the chloroplast, for even chloroplasts within the same cell may show differences as prominent as between 8.2.1 and 8.2.1a. Essentially, the picture in 8.2.1a is a negative image of that in 8.2.1. The membranes are lighter (lower density to electrons) than the material between them. Whether the thin dark line separating membrane from stroma is part of the membrane or some kind of phase boundary that has accumulated osmium is not clear.

In both 8.2.1 and in text Fig. 8.2.1a, it is possible to identify small dense granules free in the plastid stroma. These are about 200 Å in diameter, slightly smaller than the ribosomes of the cytoplasm, but nonetheless to be regarded as ribosomes (R). They are thought to be involved in the synthesis of the structural proteins and enzymes which are parts of the chloroplast itself. In their random distribution and freedom from membrane association they are reminiscent of the ribosomes in bacteria and in growing cells.

Starch grains seem to develop as oval condensations of homogenous character within the chloroplast stroma (text Fig. 8.2.1a). They lack a membrane but seem consistently to be set off by a marginal zone of low density. This latter aspect is conceivably a product of shrinkage referable to the procedures of dehydration.

As pointed out above, the grana of the chloroplast (or more specifically the thylakoids of the grana) are connected through the stroma by a simple extension of their membranes to form intergranal lamellae or "frets". The connecting membranes are clearly evident in Fig. 8.2.1. It is also true that the intracisternal space is continuous from one granum to another. Indeed a closer study of the micrographs, especially that in text Fig. 8.2.1b (arrows), establishes that continuity of this space exists between adjacent thylakoids where their margins overlap at the edge of the granum. Therefore, the thylakoids and frets are all apparently part of a continuous membrane-limited system.

It is perhaps not surprising to find that the thylakoid membranes in the grana, where they are in strict parallel array, possess properties not shared by those which run through the stroma. The intergranal lamellae are similar in density and thickness to the limiting membrane of the chloroplast. In text Fig. 8.2.1b there is an extraordinarily heavy dense line where the membranes in the grana are fused. In other preparations, as shown in 8.2.1, there is a narrow, less dense space between the membranes. On the surface of these membranes, which faces the thylakoid cisterna and lies opposite this zone of "fusion", there are periodic densities. These thicknesses are spaced about 20 nm apart and presumably represent the 9 nm particles which have been shown to be associated with the chloroplast lamellae. The exact geometrical arrangement of these particles, with which the chlorophyll is associated, and other proteins, as well as the phospholipids of the membrane, is uncertain.

In recent years, however, more and more attention has come to be focused on small particulate components of the membranes which constitute the grana. There exists some evidence that these particles, 18 nm in diameter, are an integral part of the unit membrane (text Fig. 8.2.1b). Thus when the two leaflets of the membrane are separated, as may happen in the fracture of frozen chloroplasts (plate 2.5), these particles are exposed. The most that can be said about their relation to photosynthesis is that the reactions in which light energy is converted into chemical energy take place in the membranous systems, whereas the dark reactions of the CO_2 reducing sequence take place in the stroma. The light reactions result in the formation of (a) the strong reductant, reduced nicotinamide-adenine dinucleotide phosphate, which drives the CO_2 reduction, and (b) some unknown oxidant which leads to the splitting of H_2O with the release of free O_2, and (C) adenosine triphosphate which is required for CO_2 fixation.

From the leaf tissue of timothy, *Phleum pratense* L.

Magnification ×150,000

8.2.1 and Text Figs. 8.2.1a and 8.2.1b Supplemental Reading

Arntzen, C. J., Dilley, R. A., Crane, F. L.: A comparison of chloroplast membrane surfaces visualized by freeze-etch and negative staining techniques, and ultrastructural characterization of membrane fractions obtained from digitonin-treated spinach chloroplasts. J. Cell Biol. **43**, 16–31 (1969).

Brookhaven Symp. in Bio. No. 19. Energy Conversion by the photosynthetic apparatus. Brookhaven Nat. Lab., Upton, New York (1967).

Gibbs, S. P.: Autoradiographic evidence for the *in situ* synthesis of chloroplast and mitochondrial RNA. J. Cell Sci. **3**, 327–340 (1968).

Ohad, I., Siekevitz, P., Palade, G. E.: Biogenesis of chloroplast membranes. I. Plastid dedifferentiation in a dark-grown algal mutant *(Chlamydomonas reinhardi)*. J. Cell Biol. **35**, 521–552 (1967).

— — — Biogenesis of chloroplast membranes. II. Plastid differentiation during greening of a dark-grown algal mutant *(Chlamydomonas reinhardi)*. J. Cell Biol. **35**, 553–584 (1967).

Packer, L., Murakami, S., Mehard, C. W.: Ion transport in chloroplasts and plant mitochondria. Ann. Rev. Plant Physiol. **21**, 271–304 (1970).

Paolillo, D. J., Jr., Reighard, J. A.: On the relationship between mature structure and ontogeny in the grana of chloroplasts. Canad. J. Botany **45**, 773–782 (1967).

Park, R. B.: Substructure of chloroplast lamellae. J. Cell Biol. **27**, 151–161 (1965).

San Pietro, A., Greer, F. A., Army, T. J.: Harvesting the sun. New York: Academic Press 1967.

Wehrmeyer, W., Röbbelen, G.: Räumliche Aspekte zur Membranschichtung in den Chloroplasten einer *Arabidopsis*-mutante unter Auswertung von Serienschnitten. III. Mitteilung. Über Membranbildungsprozesse in Chloroplasten. Planta (Berl.) **64**, 312–329 (1965).

Weier, T. E., Engelbrecht, A. H. P., Harrison, A., Risley, E. B.: Subunits in the membranes of chloroplasts of *Phaseolus vulgaris*, *Pisum sativum*, and *Aspidistra sp.* J. Ultrastruct. Res. **13**, 92–111 (1965).

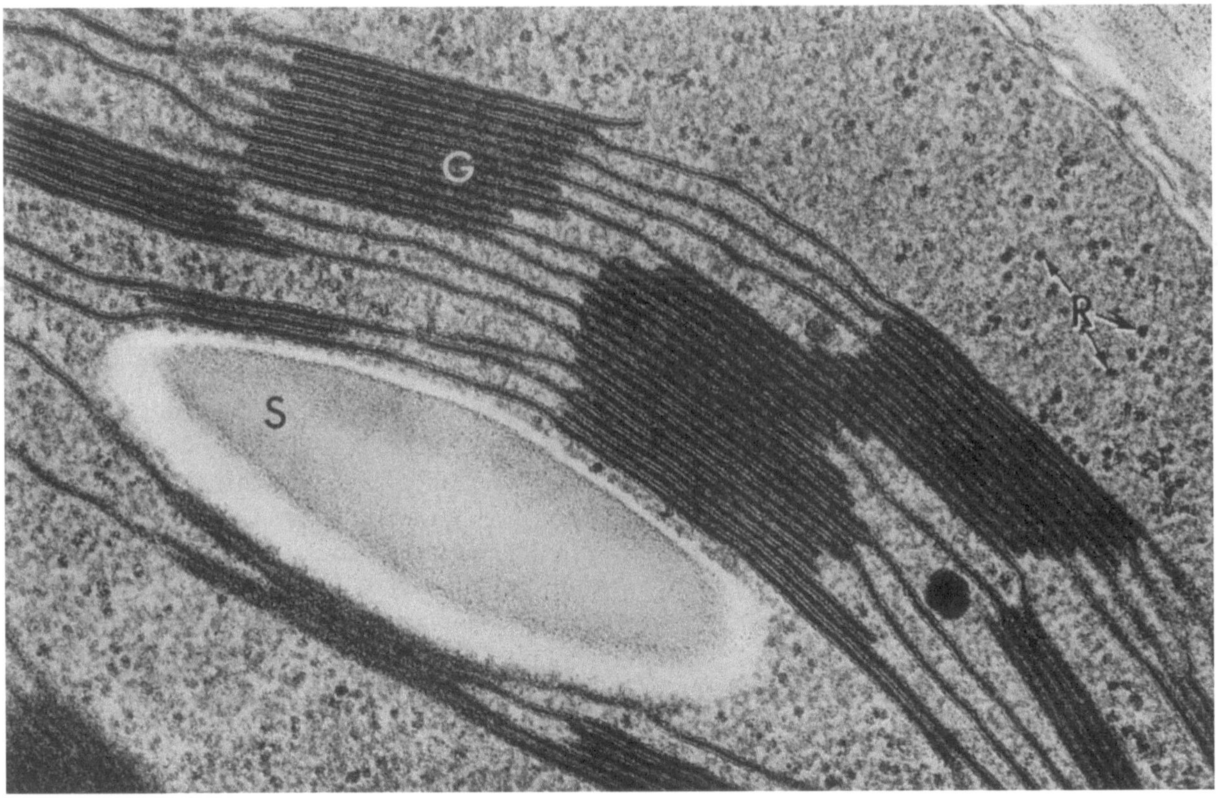
G
S
R

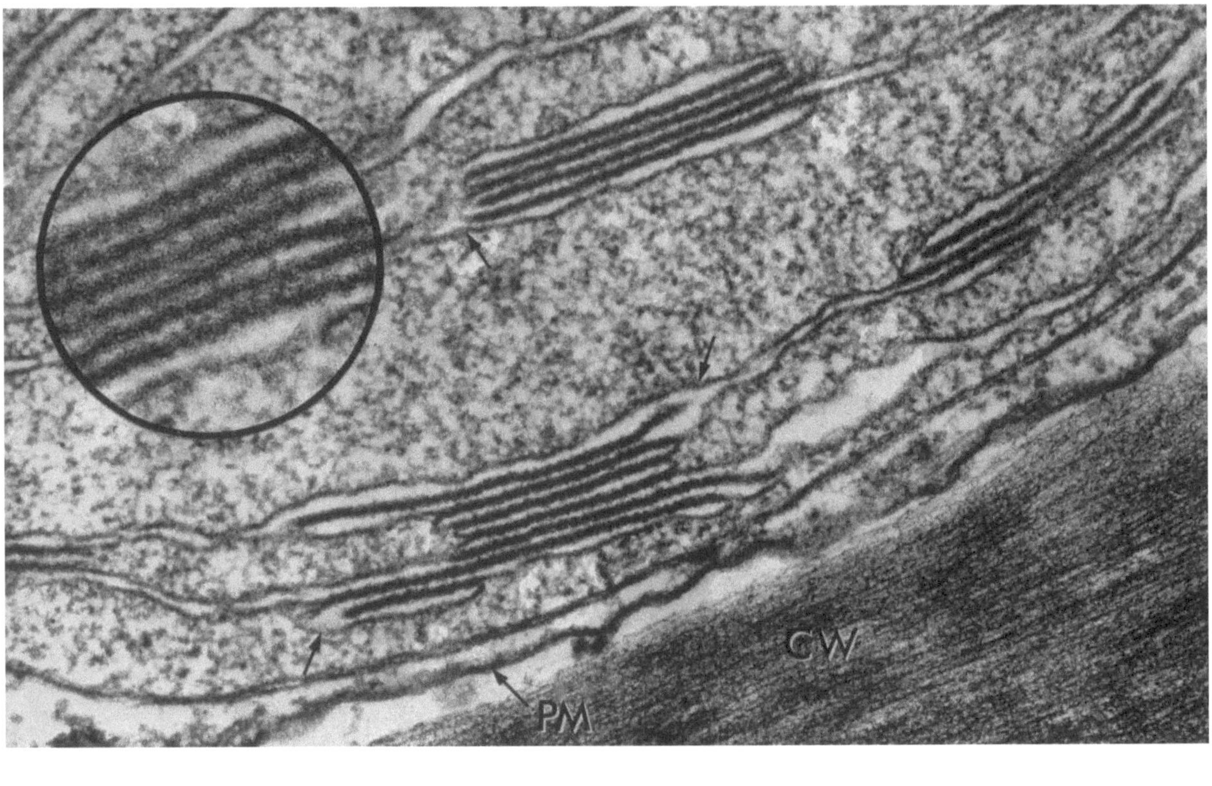
CW
PM

Text Fig. 8.2.1a

Fine Structure of Chloroplast: Anomalous Image

This depicts a vertical section through several grana in which, for some reason, the content of the thylakoid cisternae has reacted with OsO_4 to give a very dense product. Preparation procedures were identical to those used on the chloroplast shown in 8.2.1.

From the leaf tissue of *Phleum pratense* L.

Magnification ×100,000

Text Fig. 8.2.1b

Thylakoid Membranes

This micrograph of a portion of a chloroplast from a green alga clearly shows continuity of the inner space of adjacent thylakoids along the margin of the grana (arrows). In the granum at the lower center of the illustration an intergranal lemella connects the upper two thylakoid discs to the granum at the right. The fused membranes between appressed thylakoids have a wavy appearance due to the presence of regularly spaced particles. These particles are revealed in the obliquely cut granum in the insert.

From nodal cell of *Nitella* sp. Agardh.

Magnification × 82,000; insert × 160,000

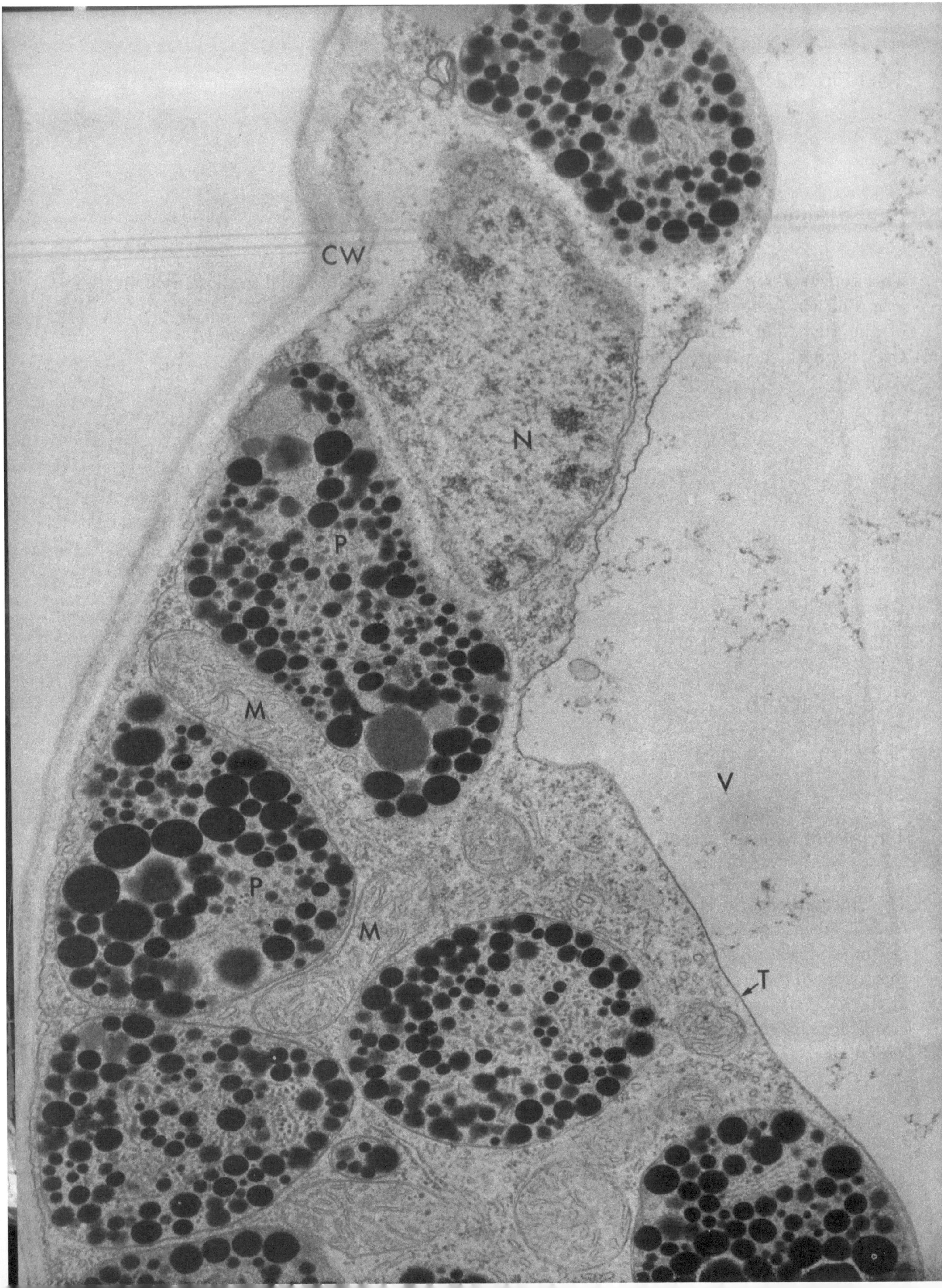
CW
N
P
M
P
M
V
T

Plate 9.1

Chromoplasts

The plant pigments, other than chlorophyll, which give flowers and leaves their color are found either in solution within the cell vacuoles or localized within special plastids, which are referred to as chromoplasts. This electron micrograph shows several chromoplasts within a portion of an epidermal cell of a petal of forsythia. The free surface of the cell, limited by a thin wall (CW) and an equally thin layer of cutin, is depicted at the left. A portion of the cell vacuole (V) with its limiting tonoplast (T) is shown at the right. The cytoplast in between contains a small end of a nucleus (N) and other inclusions, such as mitochondria (M) and ER vesicles commonly found in somatic cells.

The prominent components, however, are the plastids (P). They draw attention by virtue of their size and the conspicuous globules of floral pigment which they contain. Each chromoplast, like any chloroplast, is limited by two membranes. Within its stroma are embedded some fibrillar elements of unknown nature and a few vesicles clearly limited by membranes. The dense globular inclusions, which seem not to have a membrane, are thought to contain the yellow carotenoid pigments. These globules vary greatly in size, all the way from minute particles to bodies 500 nm in diameter. Their density, which is quite uniform, is in large measure a result of their reaction with osmium tetroxide.

The lamellar sacs, so characteristic of chloroplasts (see Plates 8.1–8.2.1), are entirely absent from these chromoplasts. This fact is even more surprising when one realizes that some types of chromoplasts are derived from previously mature and functioning chloroplasts, and that all chloroplasts contain carotenoids. The morphology of the chromoplasts illustrated here may be accounted for, in part, by a high concentration of contained carotenoids. It is less surprising that chromoplasts which develop directly from proplastids lack lamellar sacs.

It can be noted that chromoplasts also differ from chloroplasts in their size and shape. The former are generally smaller and lack the regular lens form of the chloroplast. In this micrograph they apparently conform to the space available between other organelles of the cell and seem, therefore, to possess more plasticity than do the chloroplasts. This property is also suggested by the tendency of the chromoplast to adopt the form of the crystal when the pigment within it is in crystalline form.

From an epidermal cell of a petal of *Forsythia ovata* Nakai.
Magnification $\times 25,000$

9.1 Supplemental Reading

Rosso, S. W.: The ultrastructure of chromoplast development in red tomatos. J. Ultrastruct. Res. **25**, 307–322 (1968).

Thomson, W. W.: Ultrastructural development of chromoplasts in Valencia oranges. Botan. Gaz. **127**, 133–139 (1966).

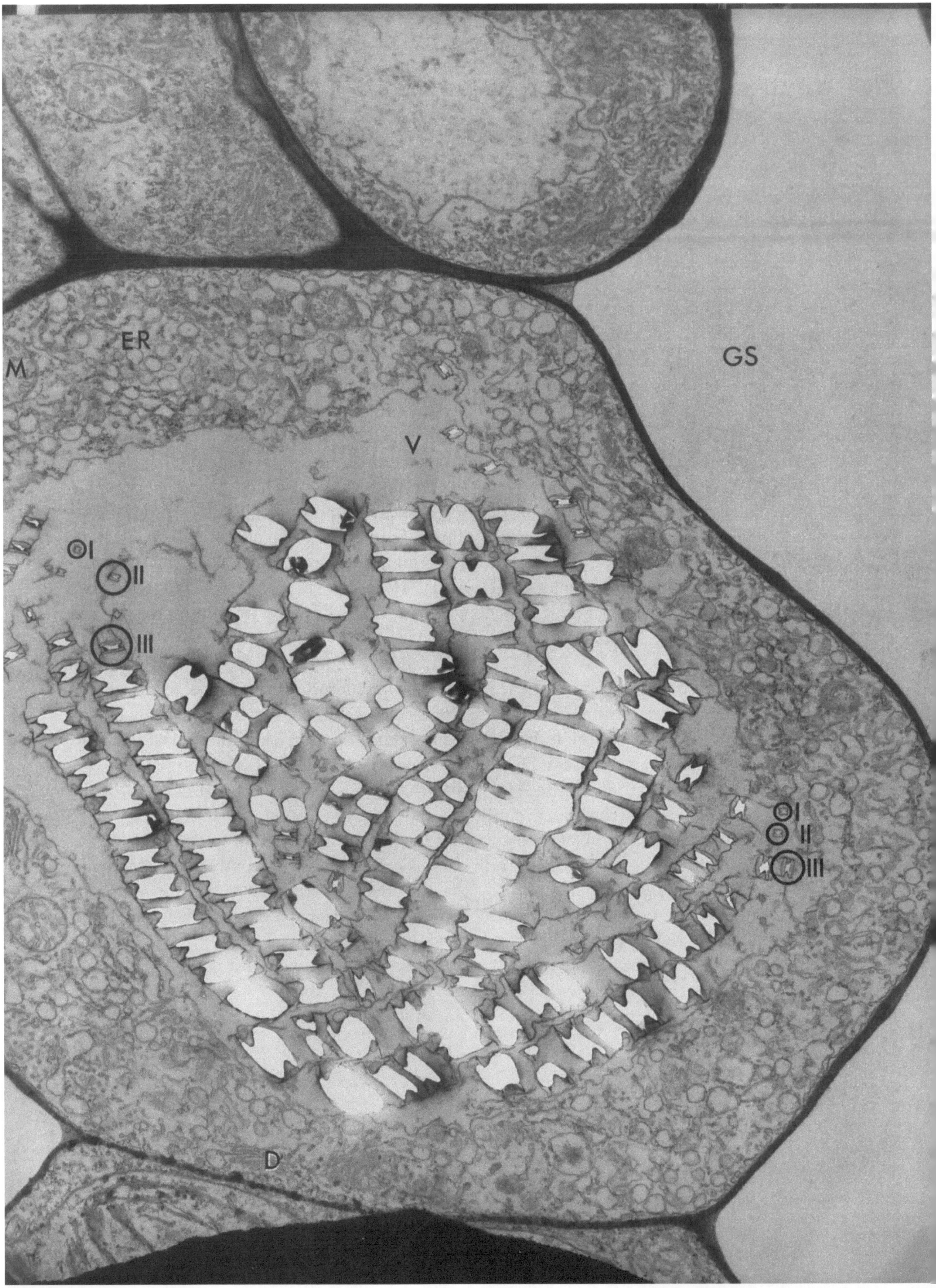
ER
M
GS
V
I
II
III
I
II
III
D

Plate 9.2

Raphid Cell

It is not uncommon for plant cells to form intracellular crystals. These may occur singly or in bundles called raphides, and usually they are contained within the vacuole. Such crystal formation is thought to be a device evolved to sequester certain metabolic wastes (e.g., oxalate), which if allowed to accumulate and diffuse freely would poison the cell or tissue. It is perhaps not surprising therefore that cells capable of storing crystals are found in all parts of the plant.

This micrograph shows such a cell from the mesophyll tissue of *Spirodela*, the duck weed. A stomatal guard cell is shown in part at the top center of the micrograph with epidermal cells to the left of it. Beneath these is the large cell whose vacuole and tonoplast seem involved in a storage function. Its central vacuole is occupied at the level of the section by crystals of calcium oxalate. Actually the crystals have dropped out or dissolved out, but their original location and shape are indicated in the negative image that remains.

The limits of this cell are easily identified by the thin dense primary wall. Just within it one sees a cytoplasm that is, as far as plant cells go, unique in containing innumerable round vesicles. Some of these, measuring 200 nm in diameter, represent parts of the endoplasmic reticulum and have ribosomes associated with their surfaces. Others seem to be smooth-surfaced vesicles of the same system. But below this level in size, identification of the vesicles is not so readily made. A number of the smaller units, not more than 300 Å in diameter, may be in the nature of pinocytotic vesicles and therefore surface-derived. Besides these as elements of the cytoplasm, mitochondria (M) and dictyosomes (D) are obvious, but the section does not include the nucleus or any plastids.

As to details of the mechanism by which the cell segregates the oxalate within its vacuole, relatively little is known, but a few points can be recognized. Note in this micrograph several small structures (encircled), the raphidosomes, which have in section the outline of an hourglass. These are membrane-limited and without visible internal structure, but it is within them that the calcium oxalate crystalizes. Several are seen just within the tonoplast. They seem to grow as the crystal within them grows and furthermore to become part of an extensive intravacuolar membrane system, which presumably serves to orient them and the crystals which develop in them.

It is an interesting and probably significant fact that the earliest signs of crystal formation appear between the two central points, marking the constriction in the profile of the raphidosome (phase II in I, II, III series). A reasonable interpretation would suggest that the membrane at these points possesses a calcium pump and that Ca^{++} concentrated therein combines with oxalate, which must be assumed to diffuse freely through these and other membranes. If the morphology depicted in the micrograph can be accepted as closely representing the native form and associations of these unusual structures, then these raphidosomes, once they start to grow, would seem to unite into a single membrane system (see phase III in picture), which probably orients them.

The origin of the raphidosomes is in doubt. There are places where they are attached to the tonoplast, but we find it difficult to judge from the evidence available whether they are a product of this ubiquitous component of the plant cell. The free edges of membrane within the population of raphidosomes inhabiting the vacuole are also most unusual, especially if they are not artifacts of the preparation procedures.

From frond tissue of duck weed, *Spirodela oligorrhiza* Kurz.
Magnification ×25,000

9.2 Supplemental Reading

Netolitsky, F.: Die Kieselkörper. Die Kalksalze als Zellinhaltskörper. In: Handbuch Pflanzen, Bd. III/I a, S. 1–127. Berlin-Nikolassee; Gebrüder Borntraeger 1929.

Olsen, C.: Absorption of calcium and formation of oxalic acid in higher green plants. C. R. Lab. Carlsberg, Ser. chim. **23**, 101 (1939).

Scott, F. M.: Distribution of calcium oxalate crystals in *Ricinus communis* in relation to tissue differentiation and presence of other ergastic substances. Botan. Gaz. **103**, 225–246 (1941).

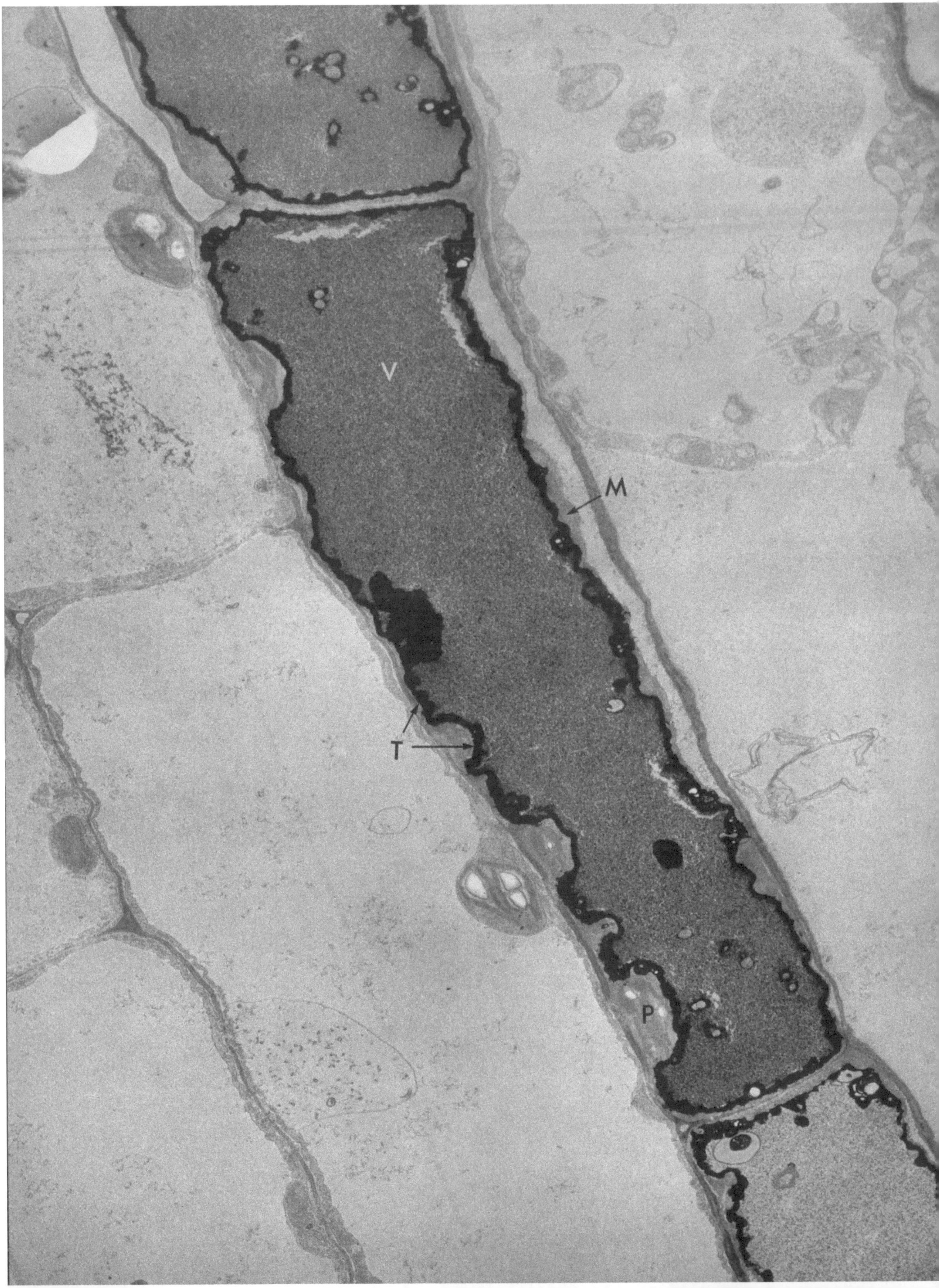
V
M
T
P

Plate 9.3

Laticifer

The laticifers make and store quantities of high molecular weight polyterpenes (latex), which are discharged if the stem is wounded and the laticifer opened. Presumably the exuded material serves to seal the lesion.

Laticifers are invariably long cells, which form files (as depicted here) and are then said to be articulated. They may develop also into long multibranched units (non-articulated) that meander randomly like the mycelium of a fungus among other cells of the stem. There are many variants and intermediates of these two forms. It has been suggested that they are related to sieve tubes, because they develop long channels and form substances similar to slime and to callose (Plate 5.3). They are common throughout the plant in certain families of angiosperms.

The cells illustrated here are immature. During ensuing differentiation their thin cross walls, or end walls, will be digested away so that all the cells and vacuoles will be interconnected to form a duct. The side walls undergo a compensatory thickening as this development progresses.

The latex itself is deposited within the vacuole. In electron micrographs it appears as a finely dispersed granular substance in the center of the vacuole (V) but is more concentrated or compacted arounds its margin. It owes its density to the fact that it reacts strongly with osmium. Information on the synthesis of latex is limited and indecisive in regard to the intracellular site of its formation. From these newer EM images one would judge, however, that the tonoplast (T) is probably involved. Beyond this, little can be said, for in these images there is no evidence of latex droplets or deposits of any kind in the thin layer of cytoplasm which surrounds the vacuole. Although the cytoplasmic layer does contain mitochondria (M) and chloroplasts (P) (made especially easy to indentify because of the contained starch), there are no obvious differences in fine structure between this and the cytoplasm of adjacent cells.

From a petiole of *Ficus religiosa* L.
Magnification ×18,000

9.3 Supplemental Reading

Mahlberg, P. G., Sabharwal, P. S.: Mitotic waves in laticifers of *Euphorbia marginata*. Science **152**, 518–519 (1966).

— — Mitosis in the non-articulated laticifer of *Euphorbia marginata*. Amer. J. Botany **54**, 465–472 (1967).

Mahlberg, P.G., Sabharwal, P. S.: Origin and early development of non-articulated laticifers in embryos of *Euphorbia marginata*. Amer. J. Botany **55**, 375–381 (1968).

Rosowski, J. R.: Laticifer morphology in the mature stem and leaf of *Euphorbia supina*. Botan. Gaz. **129**, 113–120 (1968).

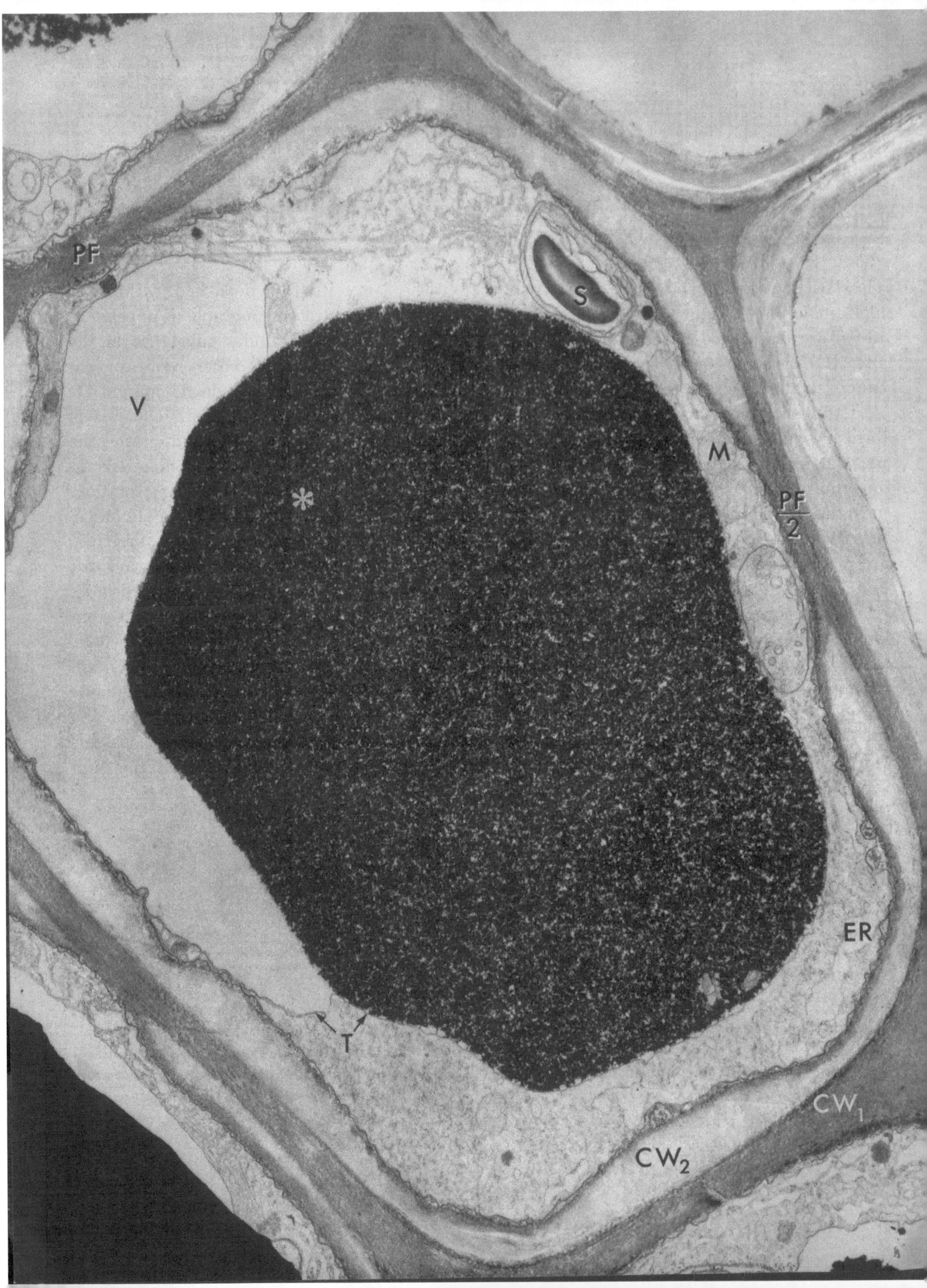

PF
S
V
M
*
PF
2
ER
T
CW1
CW2

Plate 9.4

Tannin Cell

This micrograph provides a view of a tannin cell as found among vascular and parenchymal cells in a pine needle (leaf). The dense central mass (*), which has reacted strongly with OsO_4 during fixation, represents a tannin deposit in the central vacuole.

Cells of this kind occur very widely among the higher plants and in a variety of forms and places within the plant. They may exist singly, then called idioblasts, or in characteristic groups or tissues. No general rule seems to govern the distribution of tannins, for they are found deposited extracellularly along the primary walls, dispersed in the cytoplasm or concentrated, as here, in the vacuole (V). The functions ascribed to these deposits likewise vary widely. Some hypotheses have been based on their properties as colloids, others on their antioxidant activity and still others on their capacity to discourage the invasion of other organisms and predators. It is also claimed that deposits of tannins serve as reserve materials for carbohydrate metabolism under conditions that deprive the plant of other, more usual sources.

In static images such as this there are really no clues to how the tannin is deposited. The cytoplasm surrounding the vacuole possesses the usual complement of organelles and systems, and in none is there any evidence to suggest tannin precursors. The chloroplast at the upper right contains a prominent starch grain (S). Slightly apart from it in the cytoplasm there are mitochondria (M), dilated cisternae of the ER and elsewhere dictyosomes. We can see that the cytoplasmic layer may become extremely attenuated as it is at the left in this micrograph. A few dense granules are distributed throughout the cytoplasm and seem to reside free in the ground substance. Their size (ca. 150 nm in diameter) suggests they may be glycogen deposits. Whether they may represent tannin or a precursor is not known. These or any precursor of the tannins would have to pass the tonoplast (T) in reaching the vacuole.

The protoplast is encased in secondary (CW_2) and primary (CW_1) walls. The former is the less dense and is discontinuous over pit fields. One of these (PF), a paired pit field, is included at the upper left in association with another parenchymal, possibly another tannin cell. Then at the right, an unpaired or blind pit field (PF/2) faces on a maturing tracheid. The primary, or denser appearing component of the wall, is shared with that of adjacent cells and varies less in its thickness, except where it extends out between tracheids, as at the upper right. There the primary wall is much thinner. From a parenchymal cell in needle of pine, *Pinus strobus* L. Magnification ×17,000

9.4 Supplemental Reading

Bate-Smith, E. C., Metcalfe, C. R.: Leuco-anthocyanins. III. The nature and systematic distribution of tannins in dicotyledonous plants. J. Linnean Soc. Botany **55**, 669–705 (1957).

Swain, T.: The tannins. In: Plant biochemistry, p. 552–580 (Bonner, J., and Varner, J. E., eds.). New York: Academic Press 1965.

Wardrop, A. B., Cronshaw, J.: Formation of phenolic substances in the ray parenchyma of angiosperms. Nature (Lond.) **193**, 90–92 (1962).

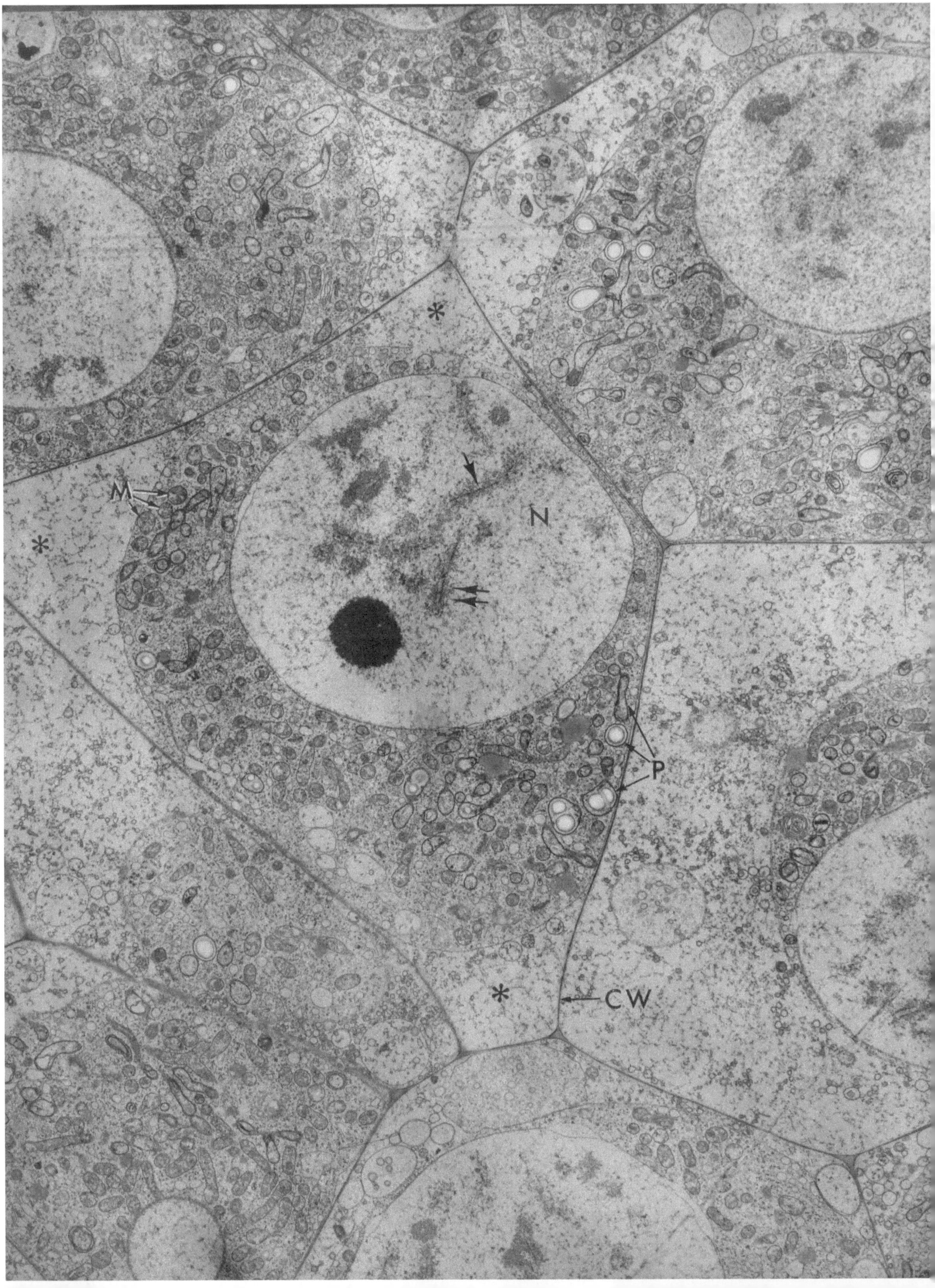

M
*
*
N
P
*
CW

Plate 10.1

Sporogenous Cells of Anther

As in other biological forms, so also in the higher plants, special cells are set aside for differentiation into reproductive units, first as spores and subsequently as definitive eggs and sperm. The earliest appearing of these are in the male parts of the flower, the anthers. They may be detected days or even weeks before the opening of the mature flower. At a time when it is hardly recognizable as the precursor of its later form, the anther, or microsporangium, develops a central mass of closely packed sporogenous cells covered by an epidermal layer, which becomes the sporangium wall. The cells of the central mass are separated from one another only by extraordinarily thin primary walls (200 Å thick), and there are no intervening extracellular spaces. The packing of the cells is analogous to that shown by a mass of soap bubbles of small size, and like the latter the average number of sides per cell approaches 14 (a tetrakaidecahedron). The majority of cell profiles evident in sections show five or six sides.

The nuclei (N) of these sporogenous cells are generally spherical and undistorted by vacuoles or other cytoplasmic components. Large vacuoles are in fact not present in this kind of cell. The envelopes surrounding the nuclei are thin and scarcely resolvable into two membranes. Pores are present in the envelope but in small numbers. The nucleoplasm, apart from the chromosomes, is remarkably transparent, probably highly hydrated, and is represented by little more than a random distribution of precipitated materials.

The meiotic chromosomes are the most noteworthy components of the nuclei at this time. In these the DNA is condensed into thread-like chromatids, which occur in pairs like lampbrush chromosomes. Homologous chromosomes are in turn aligned along their full lengths. The zone or plane of association takes on a special form known as the synaptenemal junction or complex, which is marked by arrows (single and double). Each complex, whether in cross or longitudinal section, consists of two dense lines, one on the surface of each homologue, with the pairs facing each other across a less dense (interchromosomal) zone or space about 300 nm wide. This synaptinemal association persists through the zygotene stage of prophase and ends in the separation of the homologues during the later stages of meiosis. The same sequence of events in meiosis is evident with only minor variations in other biological forms, plant and animal. Eventually each of the spore mother cells represented here will give rise to a tetrad of spores, or pollen grains, and these haploid cells and their progeny will be the microgametophytes.

One of the earliest events in the isolation of the spore mother cells from the tapetal tissue of the parent plant (sporophyte) is shown here in the withdrawal of the plasma membrane from the primary wall (see at *). The extracellular zone thus created contains numerous small vesicles, which are thought to contribute materials and possibly enzymes essential to the formation of the callose, a layer which comes to surround and thereby to isolate both the spore mother cells and the microspores. Whether the plasma membrane retains its integrity during this shrinking process is not evident.

The organelles usually found in meristematic cells can be quite readily identified here. Mitochondria (M) are small and numerous. The plastids (P), also small, frequently have starch grains that presumably are stored and used as energy for the events associated with sporogenesis. Dictyosomes are small and insignificant. There are also homogeneous granules of varying shapes, which are interpreted as protein. The endoplasmic reticulum is represented by numerous small spherical vesicles.

In withdrawing from the primary wall, these protoplasts apparently sever their plasmodesmal connections with neighboring protoplasts except at a few sites. The few plasmodesmata that remain persist for a limited period, possibly aiding in synchronizing the ensuing meiotic divisions which lead to the production of the microspores (see Plate 10.2). From the anther of the African violet, *Saintpaulia ionantha* Wendl.
Magnification ×9,100

10.1 Supplemental Reading

Echlin, P.: Pollen. Sci. Amer. **218** (No. 4), 80–90 (1968).
Matzke, E. B.: The three-dimensional shape of bubbles in foam—an analysis of the role of surface forces in three-dimensional cell shape determination. Amer. J. Botany **33**, 58–80 (1946).

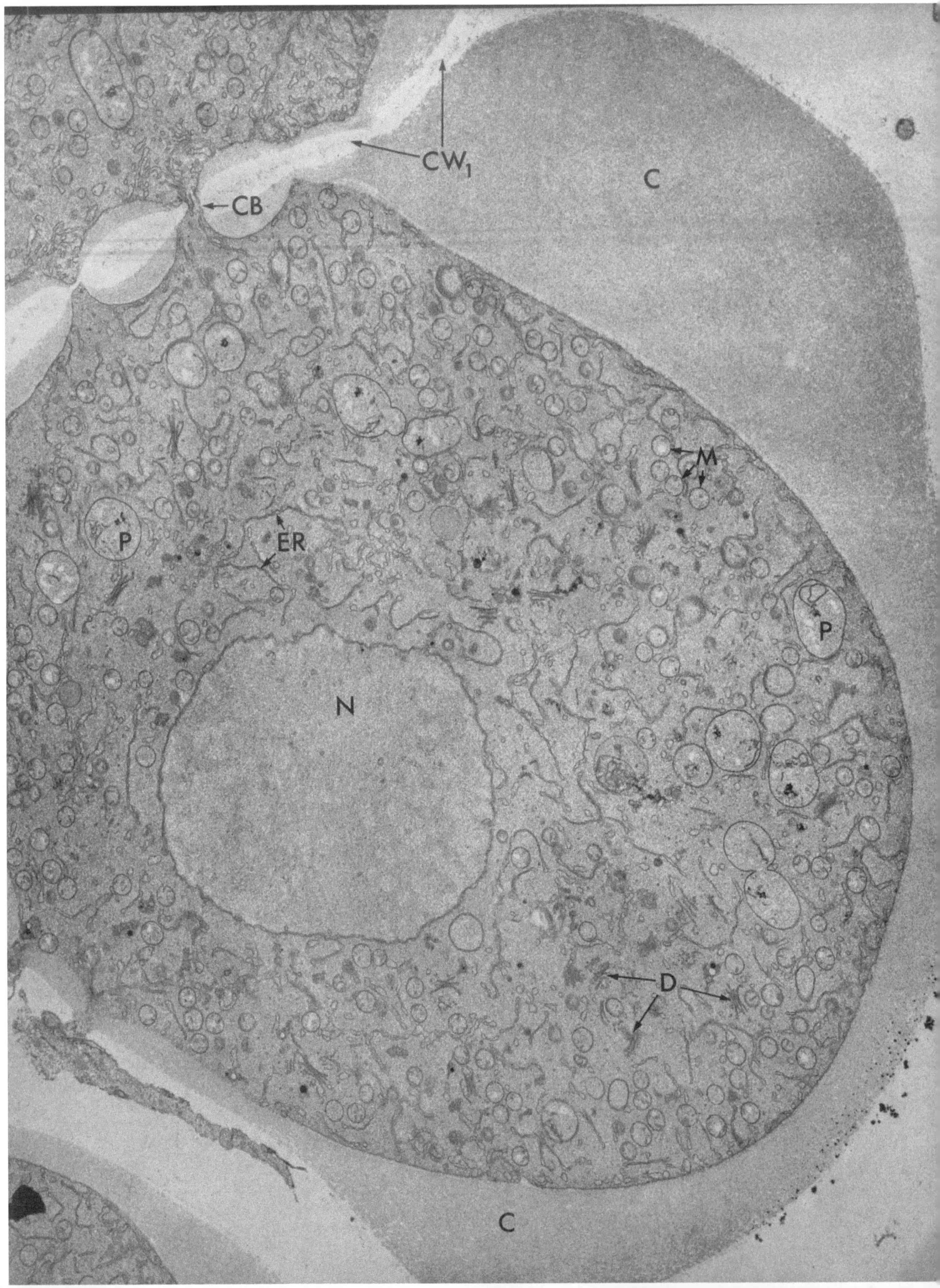
CW1
CB
C
M
ER
P
P
N
D
C

Plate 10.2

Pollen Mother Cell

After the development and segregation of sporogenous tissue (depicted in Plate 10.1), the individual spore mother cells begin to take form. One of these, showing persistent connections with adjacent cells, is pictured in this plate as it appears after fixation in potassium permanganate. The tissue was derived from the anther of *Nicotiana tabacum* L.

The major cell shown here is surrounded by a dense layer of homogeneous material called callose (C). External to it is all that remains of the thin primary wall which was present earlier (see Plate 10.1) and which has by this stage been hydrolyzed. Cytoplasmic bridges (CB) extend through the callose at scattered points to connect adjacent spore mother cells and make of the sporangium contents a syncytium. These connections are substantial in the sense that they may be as broad as 1.5 μm and may permit the free exchange of organelles between the sporogenous cells. Whether such connections represent former plasmodesmata that have expanded in girth as the primary wall has broken down, or whether they are new connections is not at once evident. Regardless of their origin they are reminiscent of connections that one finds between germ cells in many forms (animal and plant) and doubtless synchronize the meiotic divisions and to some extent the subsequent differentiations of the spores.

It is of some interest that cytoplasmic organelles occur here in greater number than in the previous image, and this is true even when comparisons are made between stages in sporogenesis within a single species. This is then a period of organelle proliferation and a general increase in cytoplasmic volume. It seems also to be a period when there is some segregation of organelles geographically within the cell. The mitochondria (M) and plastids (P) are seen to reside largely in the outer third or half of the cytoplasm leaving the perinuclear zone to elements of the ER. Dictyosomes (D) appear in greater prominence than in Plate 10.1. This segregation is thought to represent an early stage in the morphogenetic process leading to the formation of nuclear caps encountered in the tetrad spores shown in Plate 10.3.

The absence of starch from the plastids reflects only the inability of permanganate to fix it, for in other preparations fixed in osmium tetroxide and showing stages before and after (in time) that shown here, starch is present (Plates 10.1 and 10.3).

From the anther of tobacco, *Nicotiana tabacum* L. Magnification ×12,000

10.2 Supplemental Reading

Heslop-Harrison, J.: Cell walls, cell membranes and protoplasmic connections during meiosis and pollen development, p. 39–51. In: Pollen physiology and fertilization (H. F. Linskens, ed.). Amsterdam: North-Holland Publ. Co. 1964.

Spitzer, N. C.: Low resistance connections between cells in the developing anther of the lily. J. Cell Biol. **45**, 565–575 (1970).

Zamboni, L., Gondos, B.: Intercellular bridges and synchronization of germ cell differentiation during oogenesis in the rabbit. J. Cell Biol. **36**, 276–282 (1968).

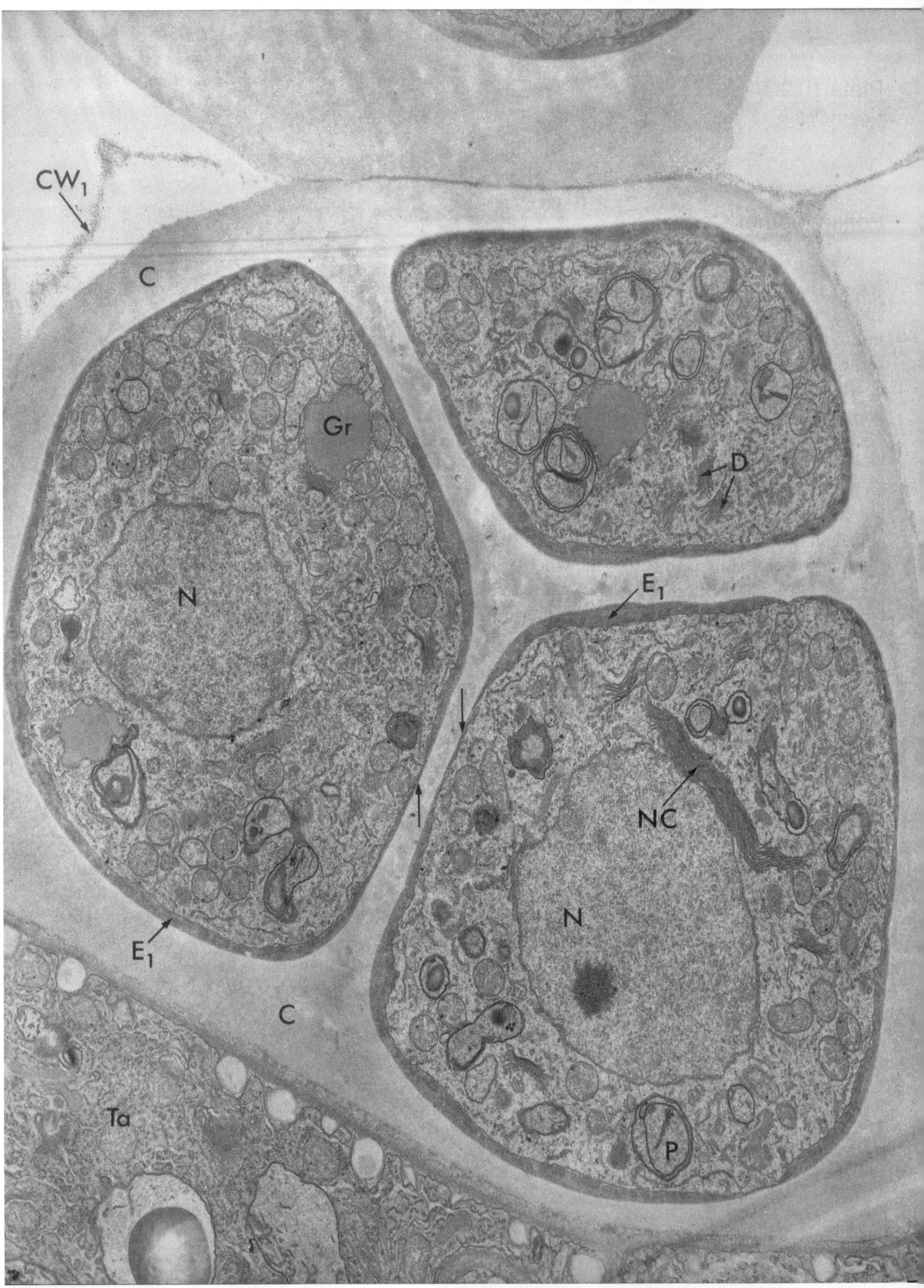
CW1
C
Gr
D
N
E1
NC
N
E1
C
Ta
P

Plate 10.3

Tetrad of Microspores

Subsequent to the stage of sporogenesis shown in Plate 10.2, the nucleus of each spore mother cell goes through two divisions of meiosis to produce the four haploid nuclei of the tetrad of microspores. Three of such a group are shown in this plate. Complete separation of the four microspores involves the development of callose (C) walls that form partitions between them. This cytokinesis is achieved by a combination of furrowing of the protoplast surface and a fusion of vesicles into which callose is rapidly deposited from the surfaces of the mother cell protoplast. The events associated with this separation are reminiscent of both plate formation (Plates 3.1 and text Fig. 3.1a), as encountered in the cytokinesis of other plant cells, and the furrowing more characteristic in the division of animal cells. The callose produced is easily distinguished by its mottled appearance from the layer which earlier surrounded the mother cell. These callose layers should not be thought of as rigid cellulose walls but rather as gelatinous, highly hydrated layers, which are easily remolded to accomodate the changing microspores. Later they are digested (see below).

The individual microspores soon acquire distinctive characteristics, especially in their cytoplasms. Organelles per unit volume of cytoplasm are extraordinarily numerous. Dictyosomes increase in prominence as the cell begins to deposit the primary exine layer (E_1). They show an increased number of cisternae and satellite vesicles characteristic of the hyperactivity associated with wall formation in other situations. The plastids gain in size. Protein granules, carried over from the early spore mother cells, are still evident, though they are probably now being used up quite rapidly in the differentiation of the pollen grains. Perhaps more striking than any other change is the appearance of the nuclear cap (NC), an extensive stack of about 10 thin cisternae. The cap is a unique disc-shaped structure, about 10 μm in diameter, closely applied to one pole of the nucleus. The presence of ribosomes on the outer cisternae of the stack relates the entire structure to the endoplasmic reticulum. Its earlier development has not been observed. Later it apparently spreads to form a complete sheath of cisternae around the generative nucleus of the pollen grain or microgametophyte (Plate 10.5). Whether this entire structure should be interpreted as a derivative of the nuclear envelope and whether it is homologous with the so-called annulate lamellae (which frequently stack up in animal egg cells) are, at this point, open questions.

It is easy in this micrograph to recognize a layer of relatively dense material just external to the plasma membrane of the protoplast, lying between it and the callose. This is the first evidence of what will eventually be part of the pollen grain wall and is appropriately referred to as the primary exine (E_1). Cellulose is probably its main component, since it resembles the appearance (after fixation) of primary walls around cells of the tapetum (Ta). Smooth on its outer surface but irregular on its inner side, the exine already is taking on the sculptured contours of its mature form. The very thin places (arrows) probably represent the future sites for the differentiation of pollen grain pores. Had more attention been given to their preservation, microtubules would probably be apparent just within the cortical zone of these small protoplasts and distributed in some meaningful relationship to the irregular thickenings that this primary exine wall (E_1) is developing.

A small portion of the tapetal tissue is included in this micrograph at the bottom (Ta). This is, of course, the sporangium wall. The cells are enclosed in a primary wall, which is contiguous with the callose. Starch grains, as storage elements, are conspicuous in these tapetal cells, and they are developing as well a prominent reticulum of ribosome-studded cisternae, which are involved at this time and later in protein synthesis. One product of this activity may be evident as small spherical globules of low density along the tapetal cell surfaces facing the callose-covered microspores. These are thought to represent packets of enzyme soon active in hydrolyzing both the callose and the primary walls limiting the tapetal cells. The same enzymes seem oddly enough to have no destructive action on the primary exine walls of the pollen grains.

From the anther of African violet, *Saintpaulia ionantha* Wendl.

Magnification ×4,300

10.3 Supplemental Reading

Angold, R. E.: The ontogeny and fine structure of the pollen grain of *Endymion non-scriptus*. Rev. Palaeobot. Palynol. **3**, 205–212 (1967).

Echlin, P., Godwin, H.: The ultrastructure and ontogeny of pollen in *Helleborus foetidus* L. II. Pollen grain development through the callose special wall stage. J. Cell Sci. **3**, 175–186 (1968).

Godwin, H.: The origin of the exine. New Phytologist **67**, 667–676 (1968).

Godwin, H., Echlin, P., Chapman, B.: The development of the pollen grain wall in *Ipomoea purpurea* L. Roth. Rev. Palaeobot. Palynol. **3**, 181–195 (1967).

Roelofsen, P. A.: The plant cell wall. Handbuch Pflanzen, Bd. 3, S. 1–336. Berlin-Nikolassee: Gebrüder Borntraeger 1959.

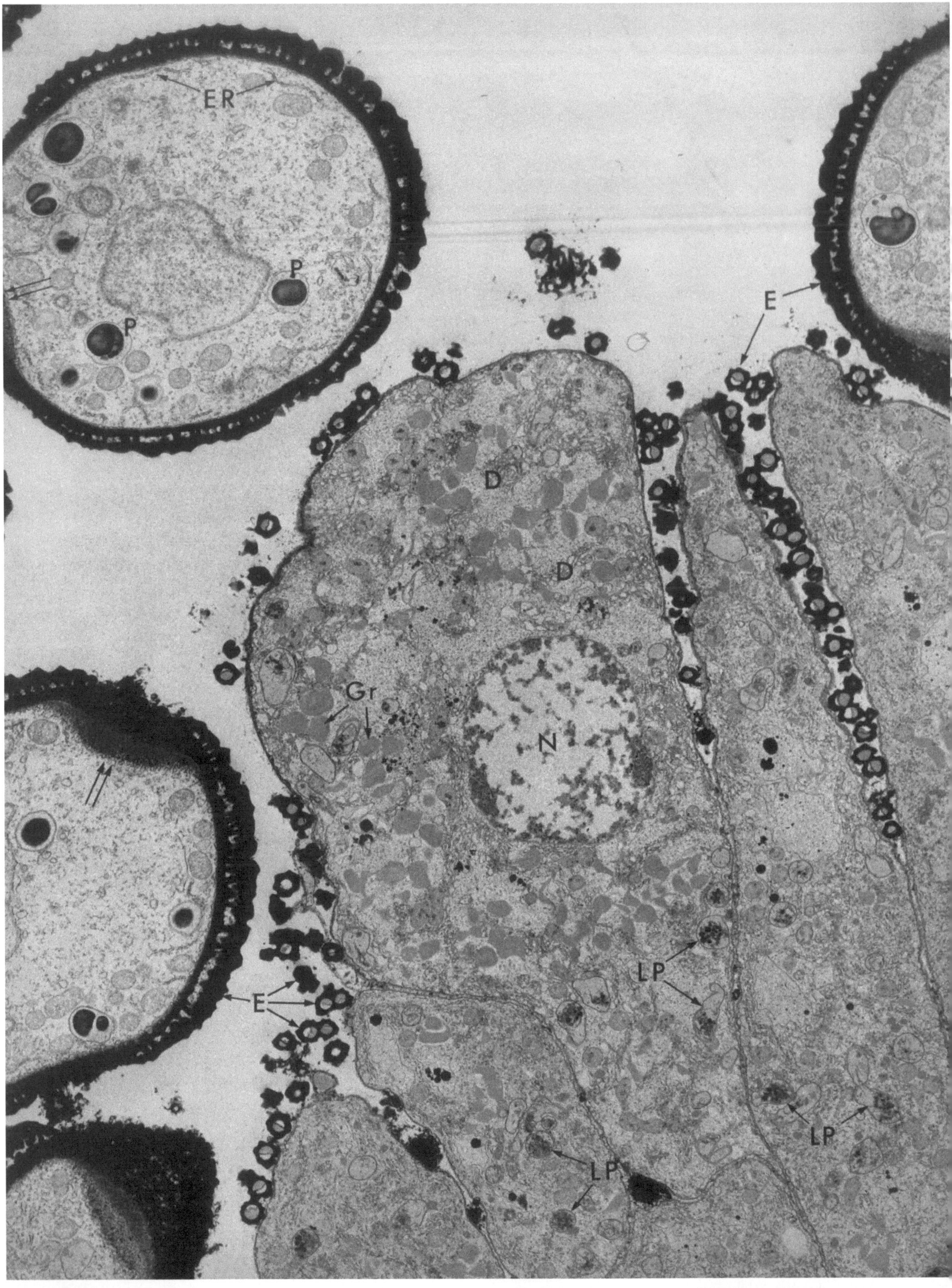
ER
P
P
E
D
D
Gr
N
LP
E
LP
LP

Plate 10.4

Tapetum and Sporopollenin Formation

At a late stage in pollen grain formation certain cells in the wall of the sporangium adopt a secretory function and engage in the manufacture of complex compounds which eventually find their way to the surfaces of the pollen grains. There they serve as protective coverings until such time as the grain reaches a suitable environment for germination. These materials, referred to collectively as *sporopollenins*, are famous for their resistance to oxidation and/or hydrolysis for extended periods, even millions of years. The molecular formula for these durable compounds is approximately $C_{90}H_{129}O_{12}(OH)_{15}$.

The cells portrayed in this micrograph include (along the upper and left-hand margins) a number of immature pollen grains. These have been freed from the callose of the earlier tetrad stages of sporogenesis and now exist as completely independent cells. Their most salient feature is their external covering, which here appears very dense, thanks primarily to their osmiophilia. The layers in these extracellular coverings, referred to as exine and intine, are shown to better advantage in other plates (10.5 and 10.6). However, it is clear even in this immature specimen, that the surface is sculptured and that its thickness and morphology change especially in regions coinciding with the germinative pores of the pollen grains (double arrows).

It is interesting to note that here, in comparison with previous stages of sporogenesis (Plate 10.3), changes in cytoplasmic content and fine structure are generally in the direction of simplification. Dictyosomes, for example, are hardly to be found. Plastids (P) are few and uniformly occupied by a large starch grain. Mitochondria are less numerous. The few remaining elements of the rough ER seem to be associated preferentially with the cortical zones of the cell as though involved in some activity going on at the cell surface. The nucleus is without distinctive morphology. Whether these features are associated with a relatively quiescent period in the life of this cell is not known. What is apparent is that it represents not more than a passing phase destined to be replaced by the greater complexity seen during gametophyte development and generative cell formation (see Plate 10.5.2).

The tapetal cells, which occupy the lower right-hand corner of the micrograph, are among the more remarkable cells of higher plants. At this stage they have only very thin walls, which in their flexibility allow for the several forms shown by these cells. The more rigid primary wall depicted in Plate 10.3 has here been removed and replaced by a layer of mucopolysaccharide, no thicker than and possibly not different from that which comprises the glycocalyx of animal cells. The absence of a wall on these cells renders their secretory products readily available to the developing pollen grains. Except indeed for a small button at the base where the primary wall persists, practically the whole surface of the cell is free (see text Fig. 10.4a). Thus the products of secretion (Ubisch bodies), evident deep in the interstices between the cells of this tissue, are able to move into the open space around the pollen grains.

Within the tapetal protoplasts there are several morphological details that are difficult to interpret. In part this is doubtless a reflection of the fact that these cells have reached the end of their differentiation and are engaged in the final stages of transforming their substance into the compounds of exine (E). The nucleus (N) appears degenerative. Both mitochondria and plastids contain small dense granules (LP) thought to be rich in lipids. Irregularly shaped granules of homogeneous nature (Gr) dominate the cytoplasm. These intermingle with remnants of an endoplasmic reticulum, which in an earlier stage of differentiation was more evenly distributed and more obviously engaged in the synthesis and segregation of proteins. Whether these granules are a product of that synthetic activity has not been determined. Whatever their nature, they are apparently discharged more or less intact from the cell to find their way into the centers of the small, spheroidal, dense-walled, exine granules (E) that are so numerous in the liquid-filled space between tapetal cells and pollen grains.

Osmiophilic substances appear in several places with respect to the tapetal cells. Within the cytoplasm there are tiny spherical droplets or granules of widely varying sizes, distributed in the cytoplasmic ground substance. These droplets seem not to bear any constant relation to the ER. Similar osmiophilic dense droplets are found between the cells, presumably where secreted. And finally they condense out as part of the extracellular granules with the low density centers (Ubisch bodies), which are absorbed onto the surfaces of the pollen grains.

From the anthers of African violet, *Saintpaulia ionantha* Wendl.

Magnification ×6,700

10.4 Supplemental Reading

Echlin, P., Godwin, H.: The ultrastructure and ontogeny of pollen in *Helleborus foetidus* L. I. The development of the tapetum and Ubisch bodies. J. Cell Sci. **3**, 161–174 (1968).

Heslop-Harrison, J.: Pollen wall development. Science **161**, 230–237 (1968).

Rowley, J. R.: Stranded arrangement of sporopollenin in the exine of microspores of *Poa annua*. Science **137**, 526–528 (1962).

— Ubisch body development in *Poa annua*. Grana Palynologica **4**, 25–36 (1963).

Rowley, J. R.: Fibrils, microtubules and lamellae in pollen grains. Rev. Palaeobot. Palynol. **3**, 213–226 (1967).

— Erdtman, G.: Sporoderm in *Populus* and *Salix*. Grana Palynologica **7**, 517–567 (1967).

— Southworth, D.: Deposition of sporopollenin on lamellae of unit membrane dimensions. Nature (Lond.) **213**, 703–704 (1967).

Tsukada, M., Rowley, J. R.: Identification of modern and fossil maize pollen. Grana Palynologica **5**, 406–412 (1964).

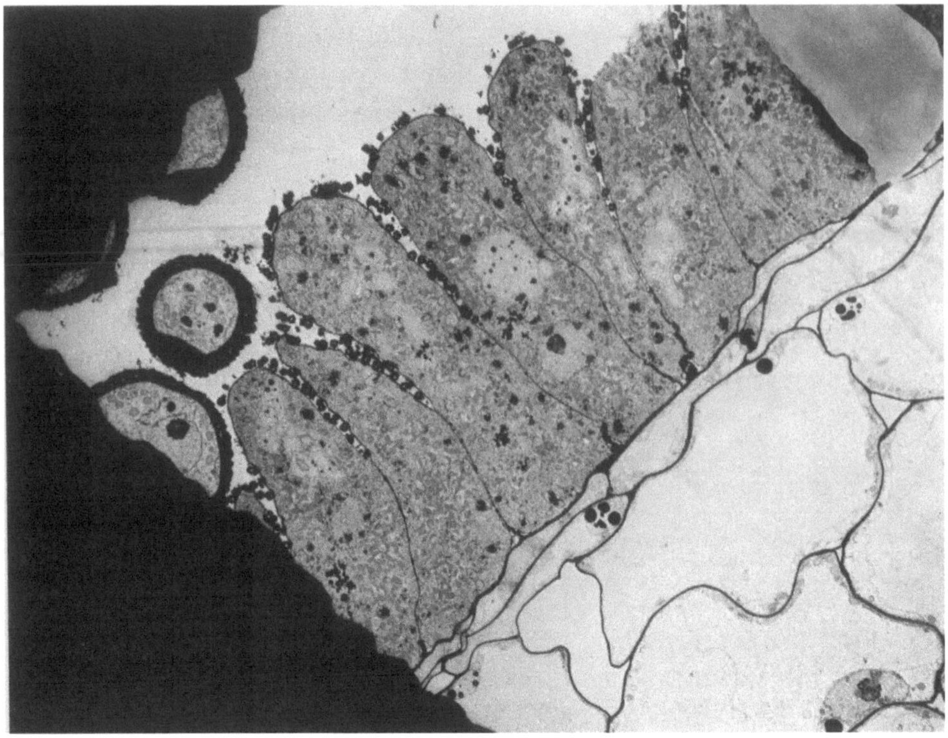

Text Fig. 10.4a

Microsporangial Wall

In this low power micrograph a row of dense columnar tapetal cells runs through the center, with the highly vacuolate cells of the anther wall at the lower right, and some maturing pollen grains at the left. (The dark areas in the left corners are bars of the grid supporting for the section). The tapetal cells are naked except for a remnant of the walls between them at their base and the walls shared with the adjacent parenchyma.

From anther of *Saintpaulia ionantha* Wendl.

Magnification $\times$ 1,400

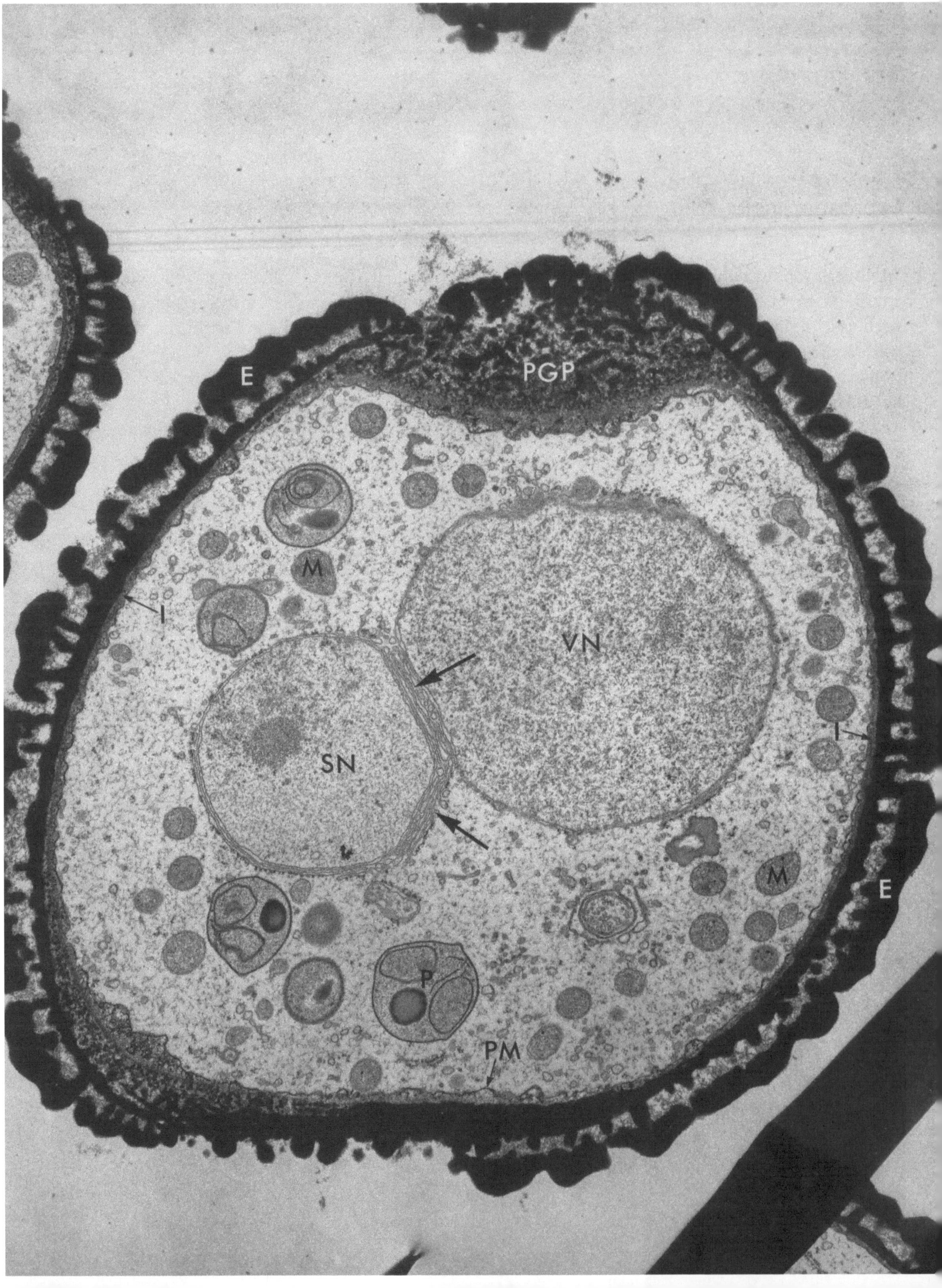

PGP
E
M
I
VN
SN
I
M
E
P
PM

Plate 10.5

Pollen Grain: Microgametophyte

The nucleus of the free and independent microspore eventually divides into a generative nucleus and a vegetative nucleus and thus initiates the development of the microgametophyte. No longer a spore, this minute haploid binucleate cell is properly called a pollen grain. As depicted in this micrograph the grain is essentially mature. It possesses a thick dense wall which, as the absence of free Ubisch granules would suggest, is complete and fully formed.

This wall, on close examination, is found to be a fairly complicated structure with an overall thickness of the order of 1.5 μm. It consists of materials supplied (a) through the synthetic activities of the microspore protoplast and deposited over it as a wall and (b) by tapetal cells, which secrete quantitites of a polyterpene (sporopollenin) that penetrates to a certain level the wall of the spore. In these electron micrographs the degree of sporopollenin penetration is believed to be indicated by the depth or presence of extremely osmiophilic material. It can be seen that this does not include the innermost layer of the wall, the intine (I), which is contiguous with the plasma membrane (PM) of the cell. According to recent observations the intine is put down by the pollen grain protoplast after the inner limits of sporopollenine penetration have been determined. The layer peripheral to the intine, called the exine (E) is thicker and more complex in its morphology. It consists of four recognized layers. Of these the outer two, the tectum and the columella, are discontinuous, whereas the inner two, together called the nexine, are continuous and less easily distinguished. Since the form of the columella can be discerned in the structure of the wall early put down around the microspore (see Plate 10.3) it would seem to follow that the less dense, finely-divided material in this layer is a residue from that earlier wall. All of these layers in the pollen grain wall continue over the regions representing pores (PGP), which are marked as slight concavities in the cell surface. In these areas, observed in section, the limits of infiltration of the sporopollenin are less sharply defined.

The markings and shapes of pollen grains, the interest of paleobotanists, reflect the uneven depositions of sporopollenin in the tectum and columella. These depositions in turn exaggerate compositional and thickness differences in the early wall of the microspore. To the extent, then that, these differences are determined by the microspore protoplast and its genetic information, the species-specific sculpturing on pollen grains is accounted for.

The microgametophyte within the pollen grain is not remarkable in its morphology. It shows here the expected two nuclei, of which one, a sperm or male nucleus (SN), is partly ensheathed by lamellar cisternae of the ER (arrows). These are probably carry-overs from the nuclear cap. The cytoplasm contains mitochondria (M), a few plastids (P) and surprisingly few other recognizable structures. Dictyosomes are not evident, ER profiles are scarce except around the generative nucleus and the ground substance shows a sparse scattering of ribosomes. But for the fact that this condition changes dramatically as the microgametophyte (with its vegetative nucleus [VN]) continues to develop (see Plate 10.5.2), this part of the pollen grain would excite little interest.

From the anther of the African violet, *Saintpaulia ionantha* Wendl.
Magnification ×18,000

10.5 Supplemental Reading

Ehrlich, H. G.: Electron microscope studies of *Saintpaulia ionantha* Wendl. pollen walls. Exp. Cell Res. **15**, 463–474 (1958).

Larson, D. A., Skvarla, J. J., Lewis, C. W., Jr.: An electron microscope study of exine stratification and fine structure. Pollen et Spores **4**, 233–246 (1962).

Rowley, J. R.: Fibrils, microtubules and lamellae in pollen grains. Rev. Palaeobot. Palynol. **3**, 213–226 (1967).

Stanley, R. G.: Physiology and uses of tree pollen. Agr. Sci. Rev. **3**, 9 (1965).

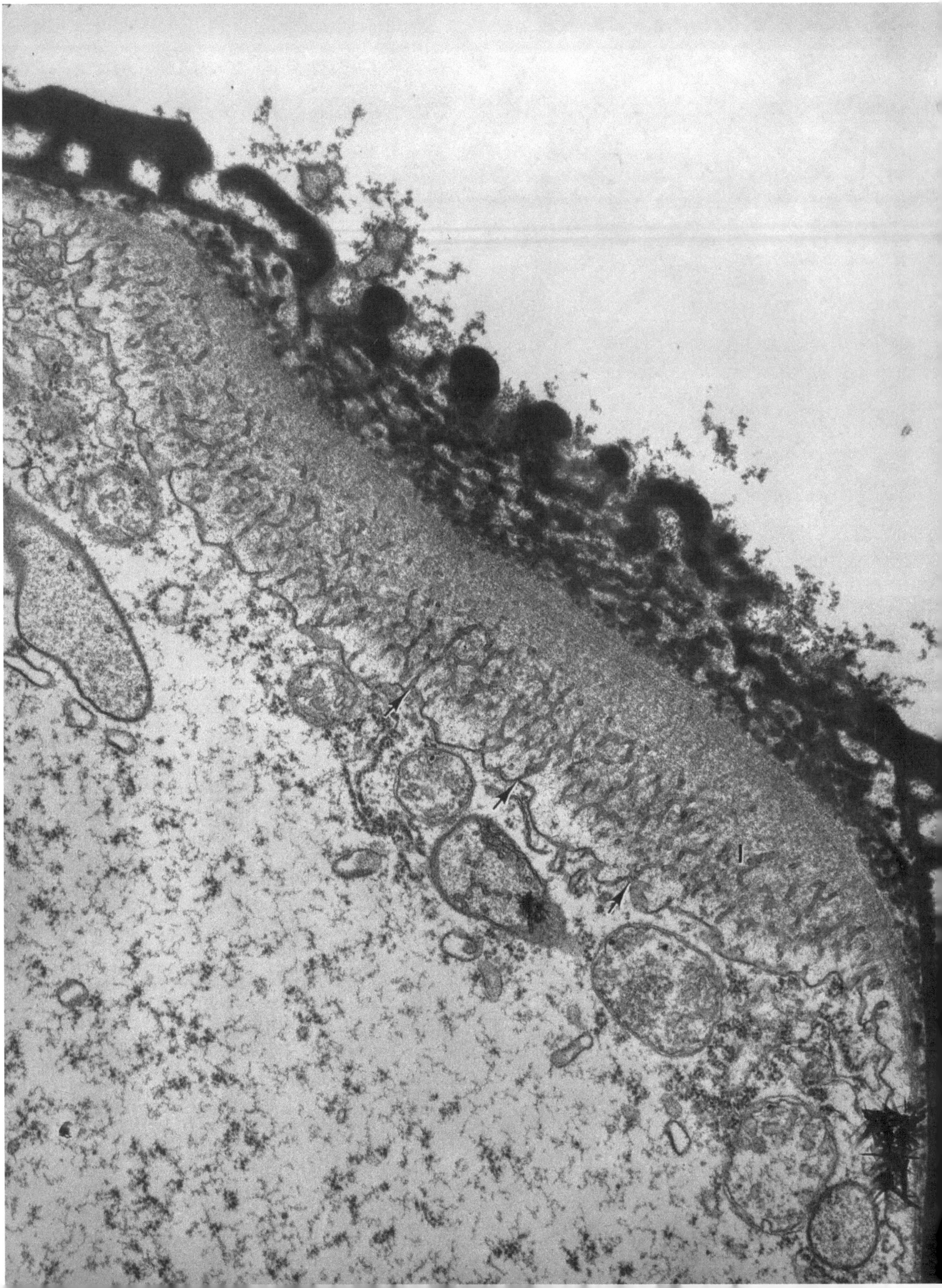

Plate 10.5.1

Pollen Grain Pore

The pollen grains of most plant species are provided with several pores, special regions in their walls through any one of which the pollen tube can emerge when the pollen grain germinates. They are the culmination of a differentiation in the microspore wall which began as far back as the tetrad stage (Plate 10.3). As well as providing for the outgrowth of the pollen tube, these special zones permit, through their greater flexibility, this otherwise rigid husk of the grain to accommodate minor volume changes resulting from variations in hydration.

As seen in this micrograph, the slightly concave pore area is sectioned vertically but somewhat off the median plane; hence the section passes obliquely through the wall. Of the five layers described previously, and quite apparent in parts of the wall peripheral to the pore, only the intine (I) is continuous over the pore as a well-defined layer. The other four layers (2 exine and 2 nexine) blend together into a single stratum of spongy character in which the sporopollenin is effectively dispersed in a finely divided latticework. Such a morphology could be expected to limit but not prevent the exchange of water and solutes between the spore cytoplasm and its environment. In this, one can recognize as well a provision for the penetration of the stimulus (boron ion), which sets off the processes leading to germination.

Internal to the wall, the protoplast at this stage in pollen grain differentiation is not changed from that illustrated in Plate 10.4. Only in the pore area is there anything of new interest, and this seems to be a further provision for germination. In this zone it can be seen that the intine is penetrated by numerous fine trabeculae, each not more than 150 Å in diameter. These appear tubular and limited by a membrane similar in thickness to that which limits the cytoplasm. Indeed where these trabeculae can be traced to the surface of the cell they can be identified as slender evaginations of that surface, continuous with the plasma membrane (arrows). It appears then that the cell surface penetrates the intine almost to its outer margin. Presumably this is a device to bring to all parts of the wall in the pore area an enzyme which when activated is capable of solubilizing the cellulose and thus opening the wall for tube emergence. It may be also that this is a form of membrane storage set up to provide for the rapid growth of the tube once germination begins.

From the anther of the African violet. *Saintpaulia ionantha* Wendl.
Magnification $\times 34{,}000$

10.5.1 Supplemental Reading

Rowley, J. R.: Stranded arrangement of sporopollenin in the exine of microspores of *Poa annua*. Science **137**, 526–528 (1962).

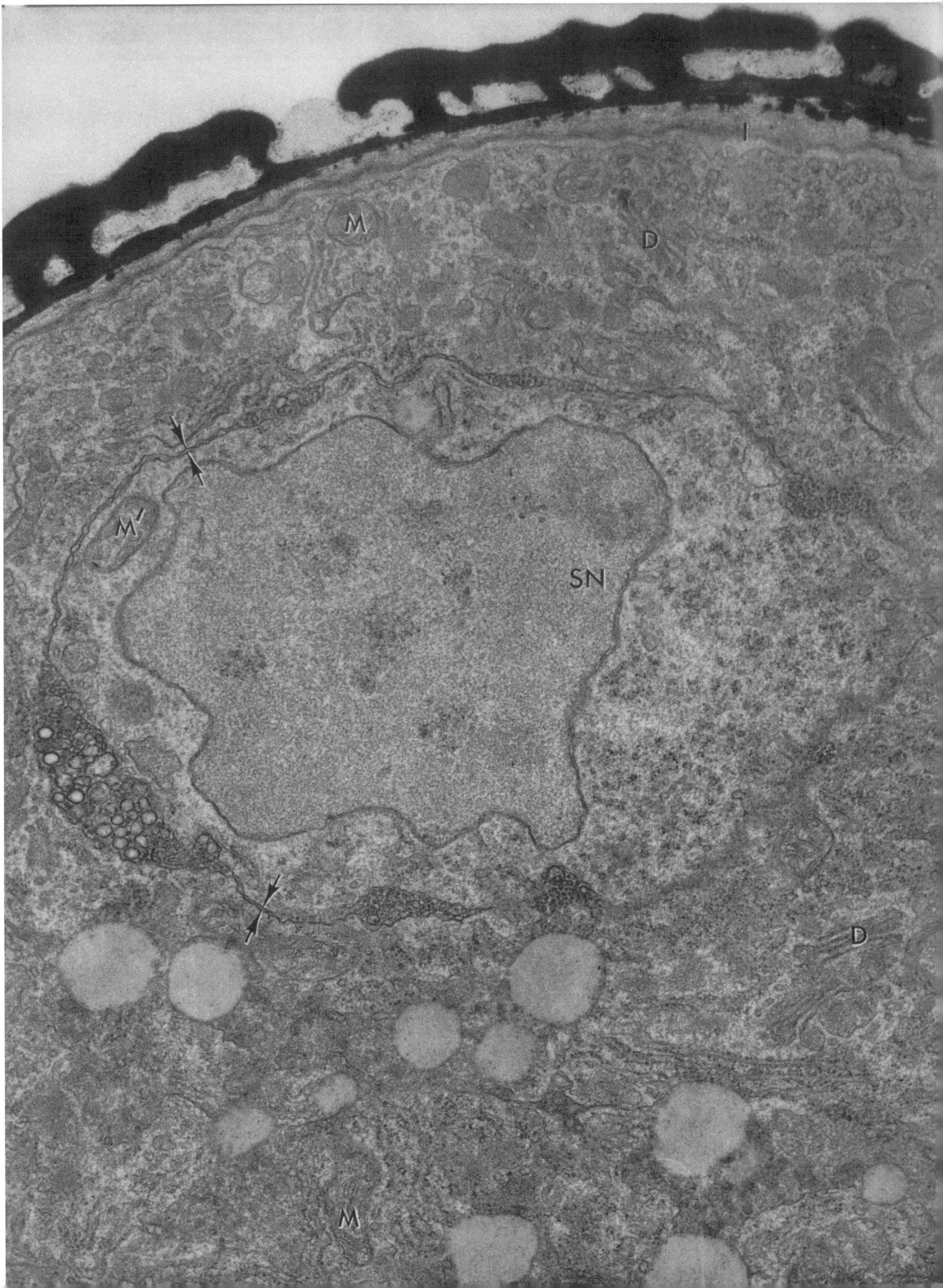
I
M
D
M'
SN
D
M

Plate 10.5.2

Generative Cell of Microgametophyte

Early in its development the microgametophyte consists of a single cytoplast with two nuclei, the generative and vegetative (Plate 10.5). The former is distinguished from the latter only by a sheath of ER cisternae apparently derived from the nuclear cap of a slightly earlier stage (Plate 10.4). Later in development a remarkable thing happens: As illustrated in this micrograph, the generative nucleus acquires its own cytoplast by a process of isolation. In this state the nucleus is carried down the pollen tube (Plate 10.5.3), simultaneously dividing once more to form two sperm nuclei. One of these fuses with the egg nucleus to start the new sporophyte generation; the other joins the polar nuclei of the megagametophyte to form the endosperm nucleus and eventually the endosperm of the seed.

The mechanism or mechanisms by which the reproductive nucleus achieves its isolation in the midst of the vegetative cytoplasm are complicated beyond current comprehension. They are doubtless related to mechanisms involved generally in determining the limits between daughter protoplasts, i.e., the location of plate formation in cytokinesis. However, a few aspects of the phenomenon do lend themselves to logical interpretation as, for example, the fusion of microvesicles in the final act in plate formation and cytoplast separation. From the morphology fortunately caught in this micrograph, one may judge that a similar event could be operating in this separation.

Obviously there are two membranes around the generative cell (arrows). One evidently belongs to the new cell; the other marks the adjacent limits of the vegetative cytoplast. The fairly large number of small vesicles caught between the membranes suggests that similar units have fused, as in plate development, to bring about the the separation. Vesicles present in excess are trapped between the two cells. The events ancillary to this incident are far less apparent. It is not evident, for example, what happens to the sheath of ER cisternae present earlier around the generative nucleus. The microvesicles involved in cell surface formation are presumably products of the dictyosomes of the vegetative cytoplasm, but no mechanism for controlling their distribution is apparent. It would seem that some unknown "influence" emanating from the generative nucleus (SN) excludes from a narrow zone of surrounding cytoplasm a clutter of organelles and inclusions, which at this time characterize the vegetative cytoplasm. The picture clearly shows the differences in the cytoplasms of the two cells. The generative cell cytoplasm contains a few mitochondria (M′) and dictyosomes (not shown here, but see Plate 10.5.3), some profiles of ER vesicles, a relatively thin scattering of ribosomes and a ground substance of low density. Plastids are conspicuously absent, a fact that fits the well known maternal inheritance of plastids established in 1909 by Correns. It seems probable therefore that the exclusion of paternal plastids from the egg is achieved in generative cell formation rather than later at the time of fertilization, as formerly assumed. Just how or why the plastids are excluded while other cytoplasmic components are included remains obscure.

The vegetative cell cytoplasm which here surrounds the generative cell has undergone some striking changes since the stage of pollen grain development depicted in Plate 10.5. In contrast to the earlier image there are now great numbers of dictyosomes (D) and mitochondria (M), and a profusion of membrane-attached ribosomes and unidentified inclusions. All these features describe unprecedented synthetic activity. Only plastids are diminished in numbers, and one is led to suggest that there has been a conversion of plastid materials to other uses. These changes, plus a further thickening of the intine layer (I) of the wall, characterize the final stages in the maturation of the pollen grain.

From the anther of African violet, *Saintpaulia ionantha* Wendl.

Magnification ×35,000

10.5.2 Supplemental Reading

Angold, R. E.: The formation of the generative cell in the pollen grain of *Endymion non-scriptus* L. J. Cell Sci. **3**, 573–578 (1968).

Brewbaker, J. L.: The distribution and phylogenetic significance of binucleate and trinucleate pollen grains in the angiosperms. Amer. J. Botany **54**, 1069–1083 (1967).

Gimenez-Martin, G., Risueno, M. C., Lopez-Saez, J. F.: Generative cell envelope in pollen grains as a secretion system, a postulate. Protoplasma (Wien) **67**, 223–235 (1969).

Heslop-Harrison, J.: Synchronous pollen mitosis and the formation of the generative cell in massulate orchids. J. Cell Sci. **3**, 457–466 (1968).

Hoefert, L. L.: Fine structure of sperm cells in pollen grains of *Beta*. Protoplasma (Wien) **68**, 237–240 (1969).

Lombardo, G., Gerola, F. M.: Cytoplasmic inheritance and ultrastructure of the male generative cell of higher plants. Planta (Berl.) **82**, 105–110 (1968).

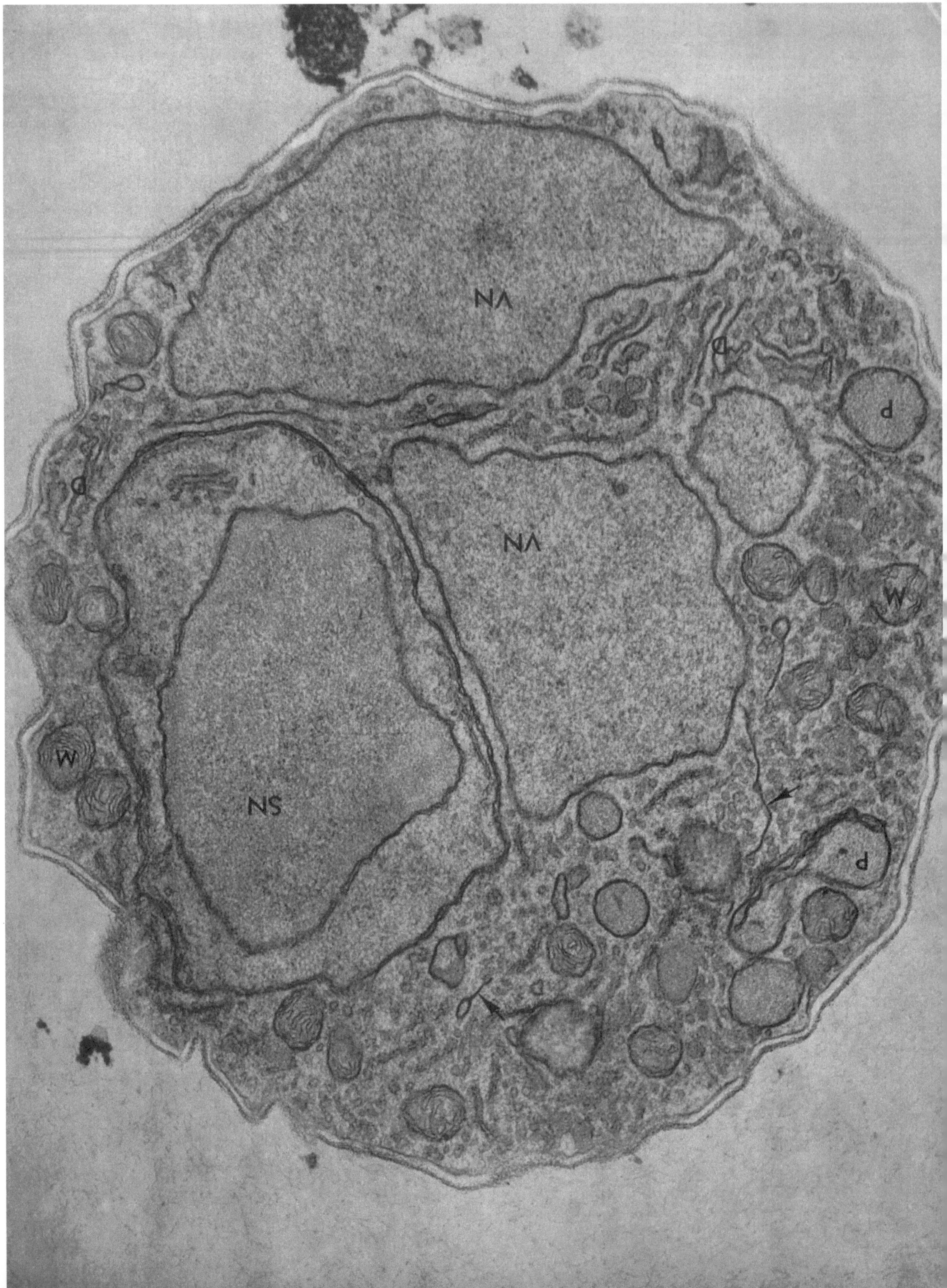

Plate 10.5.3

Pollen Tube

When mature, the pollen grain is transferred by one device or another to the receptive stigma. Subjected there to factors which stimulate germination, the grain sends out a slender tube of cytoplasm, which enters the tissue of the stigma and digests its way down the style. This growing pollen tube has the remarkable capacity not only to produce enzymes capable of solubilizing the cell walls of the style but also to metabolize the products of this digestion.

This micrograph shows a cross section of a pollen tube derived from the germination on agar of a pollen grain from *Saintpaulia*. Within the image one can readily identify parts of the multilobed vegetative nucleus (VN) and elements of the gametophyte cytoplasm such as mitochondria (M), dictyosomes (D) and plastids (P). The numerous elements of the endoplasmic reticulum are for the most part rough surfaced. Inclusions (arrows) of unknown nature (possibly part of the tonoplast) and significance appear frequently in this vegetative cytoplasm.

In the midst of these expressions of metabolic activity the sperm nucleus (SN) and its surrounding cytoplasm remain relatively simple and not obviously altered from their former appearance at a late stage in pollen grain development. The fine structure of the nucleus differs from that of the vegetative nucleus in the quantity and distribution of hetero- and euchromatin. The cytoplast, which is set off from the pollen tube cytoplasm by the expected two plasma membranes, shows a few profiles of the ER, a dictyosome and, beyond these, an otherwise homogeneous ground substance. In other sections mitochondria but no plastids can be found. It is in this form, a cell within a cell, that the reproductive cell (or cells) of the microgametophyte are carried into the egg sac.

The entire mechanism for achieving this transfer is the tube depicted here. The wall that covers its surface is very thin (about 200 Å) and is separated from the plasma membrane by a space only slightly greater in thickness than the wall. This type of wall structure is doubtless essential for such a cell extension, which has to work its way among cells of the style parenchyma.

From a pollen tube of *Saintpaulia ionantha* Wendl.

Magnification $\times 42{,}000$

10.5.3 Supplemental Reading

Beams, H. W., King, R. L.: The "negative group effect" in the pollen grains of *Vinca rosea*. J. cell. comp. Physiol. **23**, 39–46 (1944).

Bhaduri, P. H., Bhanja, P. K.: Fluorescence microscopy in the study of pollen grains and pollen tubes. Stain Technol. **37**, 351–355 (1962).

Kroh, M., Miki-Hirosige, H., Rosen, W., Loewus, F.: Incorporation of label into pollen tube walls from myoinositol-labeled *Lilium longiflorum* pistils. Plant Physiol. **45**, 92–94 (1970).

Rosen, W. G.: Studies on pollen-tube chemotropism. Amer. J. Botany **48**, 889–895 (1961).

Rosen, W. G.; Ultrastructure and physiology of pollen. Ann. Rev. Plant Physiol. **19**, 435–462 (1968).
— Gawlik, S. R.: Relation of lily pollen tube fine structure to pistil compatibility and mode of nutrition. In: Sixth Intern. Congr. Electron Microscopy, Kyoto, p. 313–314 (Uyeda, R., ed.). Tokyo: Maruzen Co. Ltd. 1966.
Rosen, W. G., Gawlik, S. R., Dashek, W. V., Siegesmund, K. A.: Fine structure and cytochemistry of *Lillium* pollen tubes. Amer. J. Botany **51**, 61–71 (1964).
Stanley, R. G., Linskens, H. F.: Protein diffusion from germinating pollen. Physiol. Plantarum (Cph.) **18**, 47–53 (1965).

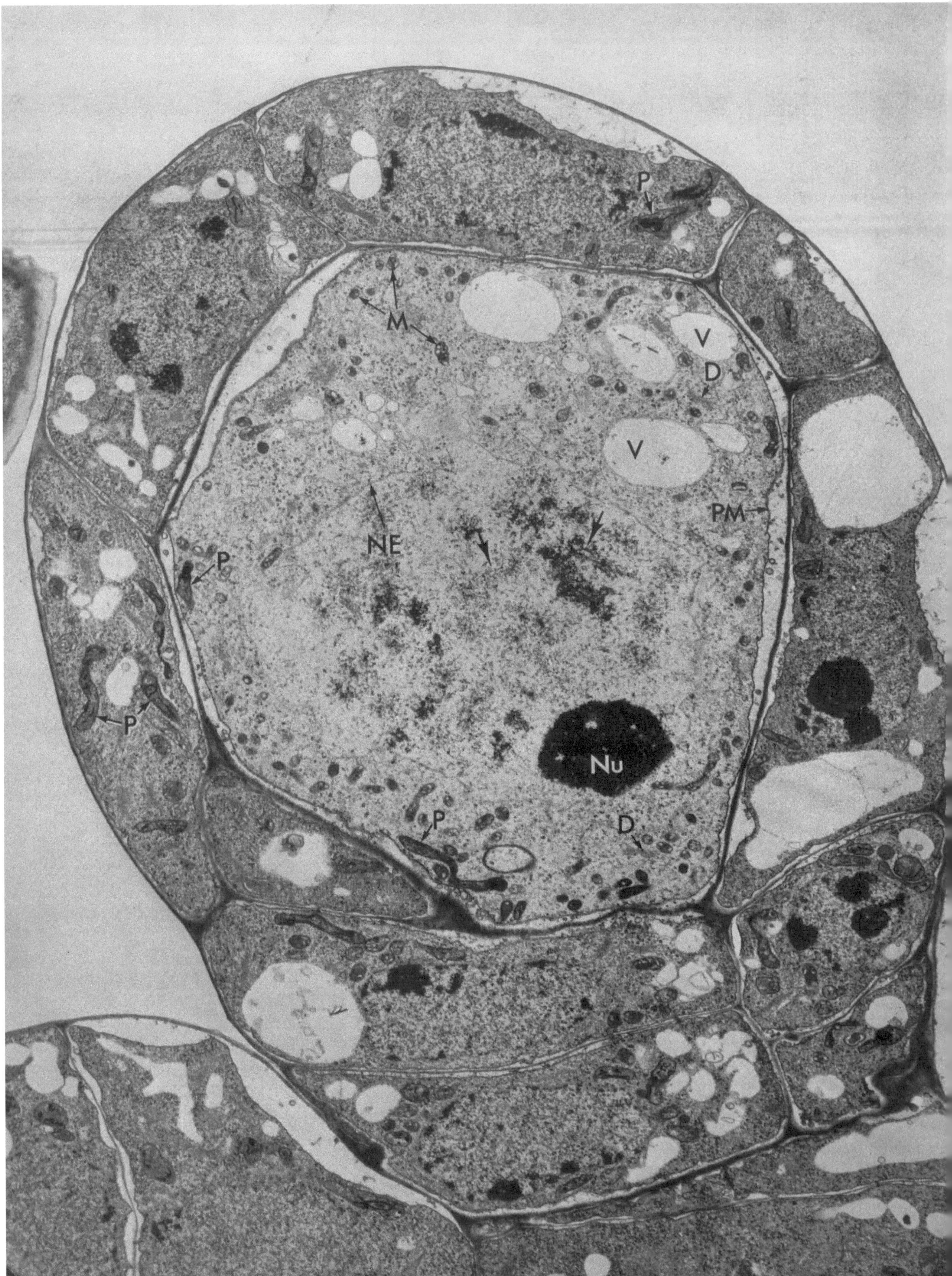

P
M
V
D
V
PM
NE
P
P
Nu
P
D

Plate 10.6

Archesporium or Megasporangium

At this early stage the megasporangium consists of two cell types. One is made up of a single layer that forms the wall of the sporangium (or nucellus) and encloses the other, the megasporocyte. Each is readily identified in this micrograph, and each is recognizably different from the other. The sporocyte will of course eventually proceed through meiosis to yield four haploid nuclei, of which three degenerate while the remaining one becomes the megaspore. The nucellar cells continue to enclose the megagametophyte and eventually give rise to the nucellar layer which surrounds the embryo. Obviously these are the initials in a long line of growth and differentiation.

The nucellar cells possess a relatively dense cytoplasm, accounted for, to a considerable degree, by a large population of ribosomes and polysomes. This condition is a characteristic one of undifferentiated and rapidly growing cells. Furthermore, such cells as these are surrounded by a thin, flexible primary wall which is penetrated by a few plasmodesmata. The latter, it should be noted, exist only between the nucellar cells and not between them and the megasporocyte. In this respect, then, the latter cell is isolated (as were the microspores in Plate 10.3) from connections which tie together other protoplasts of the sporophytic plant. The nuclei of the nucellar cells have a characteristic morphology, which includes large nucleoli and a prominent marginal distribution of heterochromatin. Cytoplasmic organelles are generally small. The chloroplasts (P) are particularly unusual. Although averaging only 0.5 μm in diameter, they may measure as much as 10–15 μm in length. The vacuoles in these same cells are also relatively small. One can recognize at this magnification only a few dictyosomes and scattered profiles of ER vesicles. The same characteristics are shared by a few cells outside the sporangial wall, i.e., in the chalaza of the sporangium.

The most prominent cell in this micrograph is, of course, the megasporocyte, which is enclosed by the sporangial wall. It is pale, less dense and presumably more hydrated than the nucellar cells. The limiting surface is represented by a thick plasma membrane (PM). In the cytoplasm mitochondria (M) and plastids (P) can be distinguished, both interspersed without evident organization among dictyosomes (D) and vacuoles (V). The plastids are small and slender (as in the nucellar cells) and seem commonly to possess a dense body in at least one end. The megasporocyte nucleus is large, irregular in outline and limited by a thin envelope (NE), which is said to possess no pores at this stage. There is a prominent nucleolus (Nu) closely associated with the nuclear envelope. Otherwise the genetic material is confined to prophase chromosomes, which at this stage (pachytene) are paired in synaptinemal complexes (arrows) (Plate 10.1). In cross section the axial threads of the homologous chromosomes appear as dense dots, and the rest as a tiny tuft of filamentous material (chromatids) around each one.

Other cells and tissues in the micrograph (at the bottom) are parts of the ovule wall and will later develop into the integuments of the seed.

From the carpel of African violet, *Saintpaulia ionantha* Wendl.

Magnification ×12,000

10.6 Supplemental Reading

Diboll, A. G.: Fine structural development of the megagematophyte of *Zea mays* following fertilization. Amer. J. Botany **55**, 797–806 (1968).

Godineau, J.-C.: Ultrastructure des différents tissus de l'ovule du *Crepis tectorum* L. au moment de la prophase meiotique. Données sur le cytoplasme de la cellule-mère de mégaspores. C. R. Soc. Biol. (Paris) **266**, 1008 (1968).

Israel, H. W.: Sagawa, Y.: Post-pollination ovule development in *Dendrobium* orchids. II. Fine structure of the nucellar and archesporial phases. Caryologia **17**, 301–316 (1964).

— — Post-pollination ovule development in *Dendrobium* orchids. III. Fine structure of meiotic prophase I. Caryologia **18**, 15–34 (1965).

Jensen, W. A., Fisher, D. B.: Cotton embryogenesis: the entrance and discharge of the pollen tube in the embryo sac. Planta (Berl.) **78**, 158–183 (1968).

Jensen, W. A.: Cotton embryogenesis. The zygote. Planta (Berl.) **79**, 346–366 (1968).

Schulz, Sister R., Jensen, W. A.: *Capsella* embryogenesis: The egg, zygote and young embryo. Amer. J. Botany **55**, 807–819 (1968).

— — *Capsella* embryogenesis: The early embryo. J. Ultrastruct. Res. **22**, 376–392 (1968).

Woodcock, C. L. F., Bell, P. R.: Features of the ultrastructure of the female gametophyte of *Myosurus minimus*. J. Ultrastruct. Res. **22**, 546–563 (1968).

MIX
Papier aus verantwortungsvollen Quellen
Paper from responsible sources
FSC® C105338

If you have any concerns about our products,
you can contact us on
ProductSafety@springernature.com

In case Publisher is established outside the EU,
the EU authorized representative is:
Springer Nature Customer Service Center GmbH
Europaplatz 3, 69115 Heidelberg, Germany

Printed by Libri Plureos GmbH
in Hamburg, Germany